Eco-Social Policy in the OECD and the World Bank

Robin Schulze Waltrup

Eco-Social Policy in the OECD and the World Bank

Global Governance and Social-Ecological Transformation

palgrave
macmillan

Robin Schulze Waltrup
Bielefeld University
Bielefeld, Germany

ISBN 978-3-031-95428-3 ISBN 978-3-031-95429-0 (eBook)
https://doi.org/10.1007/978-3-031-95429-0

Cover Credit: Hiroshi Watanabe/Getty Images

This Palgrave Macmillan imprint is published by the registered company Springer Nature Switzerland AG.
The registered company address is: Gewerbestrasse 11, 6330 Cham, Switzerland

Acknowledgement

Science is, without a doubt, a collaborative project, and the book would not have been possible without the assistance and support I received from so many. I express my gratitude to all that have supported me on this journey. First and foremost, I would like to extend my appreciation to Alexandra Kaasch and Milena Büchs. Since my time as a student assistant for Alexandra Kaasch, she has not only provided me with the opportunity to jump into the water of academia but also supported me in navigating this water's sometimes rough conditions. Her constant encouragement and implicit and explicit insights into the workings of academia have played a crucial role in shaping me as a member of the academic community, and I am genuinely grateful for the support she has provided me throughout the years. I would also like to express my sincere gratitude to Milena Büchs for her consistent guidance, expertise and for facilitating my research visit to the Sustainability Research Institute of the University of Leeds. The stimulating discussions I have had with her and the other members of the institute have been essential in finalising this research.

Additionally, I extend my gratitude to my colleagues at the Research Training Group "World Politics" at Bielefeld University for cultivating an intellectually stimulating academic environment throughout my doctoral studies. Specifically, I would like to thank Mathias Albert, Cansu Erdoğan, Jonas Fritzler, Malte Neuwinger and Martin Petzke for providing insightful comments that greatly contributed to the conceptualisation of this book's research.

Moreover, I would like to express my appreciation to John Berten and Madelaine Moore for their crucial input on my work. I am also grateful to my colleagues in the early-career researchers group of the Sustainable Welfare and Eco-Social Policy Network for all the fun discussions we have had.

Lastly, I am profoundly thankful to my loved ones, family and friends for their support and belief in my abilities. Their encouragement and understanding have been a constant source of motivation, providing me with a strong support network. Thank you to all of you!

Contents

Abbreviations

APEC	Asia-Pacific Economic Cooperation
BCA	Border Carbon Adjustments
BRICS	Brasil, Russia, India, China, South Africa
CBA	Consumption-based accounting
CCS	Carbon capture storage
CDA	Critical discourse analysis
DFI	Development finance institution
ESCP	Environmental and Social Commitment Plan
ESF	Environmental and Social Framework
ESG	Environmental, Social and Governance
ESS	Environmental and Social Standard
EPR	Environmental Performance Review
EU	European Union
EV	Electronic vehicle
FCS	Fragile and conflict-affected states
GDP	Gross Domestic Product
GHG	Greenhouse gas
GVC	Global value chain
HOUR	Time-Based
IDA	International Development Association
ILO	International Labour Organisation
IO	International Organisation
LETS	Local Exchange Trading Systems or Schemes
NAEC	New Approaches to Economic Challenges
NPM	New Public Management
OECD	Organisation for Economic Co-operation and Development

PCT	Personal carbon trading
PISA	Programme for International Student Assessment
PPP	Polluter Pays Principle
SME	Small and medium-sized enterprises
SRI	Sustainable and responsible investment
UNEP	United Nations Environment Programme
WDR	World Development Report
WISE	Centre on Well-being, Inclusion, Sustainability and Equal Opportunity

CHAPTER 1

Introduction: Towards Global Eco-Social Policy

Environment and development are not separate challenges; they are inexorably linked. Development cannot subsist upon a deteriorating environmental resource base; the environment cannot be protected when growth leaves out of account the costs of environmental destruction. These problems cannot be treated separately by fragmented institutions and policies. They are linked in a complex system of cause and effect.
—*Report of the World Commission on Environment and Development: Our Common Future (1987)*

[G]lobal social policy research and practice have to take account of the planet's ecological crisis and natural resource depletion.
—*Bob Deacon (2014 in "Toward a Transformative Global Social Policy")*

These quotes emphasise the importance of adopting an integrated perspective when formulating policies to promote environmental and social progress at the global level. Historically, ecological protection and social progress have often been seen and practised as conflicting or mutually exclusive (Fanning et al., 2022). In recent years, sustainable welfare and eco-social policy research has emerged that focuses on the tension between ecological degradation on the one hand and environmental protection and social welfare on the other. Policies aimed at bringing human civilisation

R. Schulze Waltrup, *Eco-Social Policy in the OECD and the World Bank*, https://doi.org/10.1007/978-3-031-95429-0_1

back into alignment with planetary boundaries have a significant impact on how we live, as well as on social progress and welfare, and there are many potential tensions that may arise between the behavioural and structural change necessary to reverse ecological degradation and welfare systems and labour markets (Budowski & Künzler, 2022; Fanning et al., 2020; Hirvilammi & Koch, 2020; Lamb et al., 2020; Markkanen & Anger-Kraavi, 2019). Existing welfare systems struggle to hold the balance in interactions between natural and social systems. In general, the better a country performs in meeting basic human needs, the more likely it is to exceed planetary limits. Wealthier nations tend to surpass environmental thresholds while they meet most human needs, while many low-income countries stay within these boundaries but fall short in addressing these needs for their populations (Fanning et al., 2022; O'Neill et al., 2018).

Eco-social policy research has emerged to address this tension within existing welfare systems at the policy level, insofar as it investigates processes intended to integrate environmental and social policy. Its object is the implementation of social policy that does not accelerate ecological pressures or of environmental policy that does not undermine welfare or exacerbate inequality. Eco-social policies aim instead to turn potential trade-offs between these policy fields into synergies (Mandelli, 2022; Schulze Waltrup, 2025a). Despite the global concern to create welfare systems that moderate the relationship between natural and social systems efficiently, this is the first book that provides an explicit eco-social policy assessment of international organisations (IOs)[1] as global policy actors. Most eco-social policy research and the closely related research into sustainable welfare have focused on formulating and studying particular eco-social policy proposals (e.g. Bohnenberger, 2020; Koch, 2022) and have investigated eco-social policy integration at the level of (European) nation-states (e.g. Bridgen, 2023; Hirvilammi et al., 2023; Mandelli, 2025; Zimmermann & Graziano, 2020) and of the European Union (EU) (e.g. Cotta, 2024; Laruffa & Nullmeier, 2025; Sabato & Mandelli, 2024; Schulze Waltrup et al., 2025) but fall short in understanding eco-social policy integration at the global level of IOs (but see Cigna et al., 2023; Schulze Waltrup et al., 2023). While these studies focusing on the European world-region level undeniably hold crucial value in terms of understanding processes of eco-social policy integration, the literature at

[1] This book follows the common practice of using the term "international organisation" to mean an "intergovernmental organisation".

large faces the danger of reinforcing methodological nationalism, and thus overlooking crucial global or transnational dynamics when studying these processes (Greer et al., 2020).

Even though IOs are important actors in providing policy-making frameworks to national policy-makers worldwide, we still lack knowledge of whereabouts in global governance organisations the processes and discourses that lead to the establishment of an integrated perspective take place and to what degree the process of policy-merging challenges the existing and incumbent policy notions that are often responsible for ecological and social crises. This lack of knowledge is problematic if we want to understand the potential of IOs, such as the Organisation for Economic Co-operation and Development (OECD) and the World Bank, that work closely with national policy-makers, to drive systemic change in the direction of a social-ecological transformation (see Sect. 7.2). The OECD and World Bank also drive important policy developments in the global political economy that usually reflect a hierarchical form of policy integration that works to the advantage of economic growth and conceives the environmental and social sphere as being in service to growth. Local policy-making (from municipalities to nation-states) often has to relate to these global pressures. Understanding how environmental and social policy integration occurs at the global level is, therefore, essential for identifying potential changes to capitalist dynamics, which can have an impact on all levels of policy-making. With such an understanding, it is possible to see how crises in these areas are used to reposition capitalism as the dominant logic of policy-making, both locally and globally (Schulze Waltrup et al., 2023). That is why this study provides an assessment of the OECD and the World Bank from an eco-social policy perspective. It specifically studies these two organisations because both are crucial policy actors within global governance that have a certain hegemonic status and therefore signal important policy developments on a world political scale, potentially providing policy frameworks that are perceived as appropriate for national governments worldwide. They are key actors in knowledge production and the construction of what are perceived as global policy models and best practices (De Francesco, 2014; De Francesco & Guaschino, 2019; Mahon & McBride, 2009; Orenstein, 2008; Stone, 2003). Furthermore, both organisations have a history of driving environmental and social policy and have also started to explicitly connect both policy fields (Kaasch & Schulze Waltrup, 2021; Di Leva, 2021; Hynes, 2021). But there are crucial differences between them. The World Bank focuses on driving development in the Global South (Hammer, 2013), while the OECD's particular focus is on instigating growth in its Northern member countries (Schmelzer, 2016). The World Bank is characterised by stronger connections to American knowledge networks (Wade, 2002); the OECD draws more heavily on European ones (Deacon, 2007). As comparison of the two organisations suggests, these differences result in further differences regarding each organisation's understanding of eco-social policy (Chap. 6).

This book, then, examines the following interconnected research questions:

1. To what extent have eco-social policy discourses emerged in the OECD and the World Bank?
2. What kind of eco-social policy discourses are represented in them?
3. Are the eco-social policy discourses identified driving a social-ecological transformation that rejects certain capitalist modes of production, or are they rather stabilising existing capitalist modes of production?
4. How do these eco-social policy discourses relate to the development of global policy paradigms? Is there a relatively unified "global eco-social policy paradigm"?
5. To what extent are the eco-social policy discourses characterised by struggles and conflicts between different policy perspectives, and which policy perspectives are being silenced?

By conducting a comprehensive analysis of the eco-social policy discourses of the World Bank and the OECD, two influential global policy actors, this study contributes to the development of a research field of global eco-social policy. Eco-social policy research mainly emerges from the tradition of social policy research. While it is not surprising that research on global social policy has long emphasised the role of IOs in shaping social policy discourses (Béland & Orenstein, 2013; Deacon, 2007; Martens et al., 2021) and the transnational governance structures through which actors of different levels diffuse their policy prescriptions (Kaasch & Martens, 2015), there has been a recent surge of interest among academics and practitioners within the field in the intersection and potential conflicts between environmental and social policies (Kaasch & Schulze Waltrup, 2021; Lakeman, 2021). Although some attempts have been made within global social policy research to highlight the crucial impact of environmental degradation and climate change on social policy issues and vice versa (Deeming, 2021; Gough, 2013; St. Clair & Lawson 2014), there has hardly been any empirical research let alone a book on global eco-social policy (Köhler, 2016; Lakeman, 2021; Schulze Waltrup et al., 2023). Basic knowledge of where and to what degree we can observe a merging of environmental and social policy within global policy actors is limited. Discussions on global eco-social policy have mostly been focused on raising awareness of environmental policy considerations and how they interact with global social policy arrangements. Indeed, Smyth and Deeming (2018, p. 325) claim that environmental sustainability is the biggest issue for global social policy (next to the future of work). Having that in mind, it comes as a surprise that popular global social policy notions fail to deal with environmental policy considerations. Social investment and inclusive growth, for example, have been discussed recently and framed as new leitmotifs within global social policy (Deeming & Smyth, 2018). However, they reproduce and strengthen notions of economic growth without problematising the underlying idea of productivism and the detrimental consequences of economic growth in regard to transgression of planetary boundaries. For example, the social investment perspective highlights the productive value of social policy measures in regard to supporting the growth of the economy, while inclusive growth puts economic growth at its centre with respect to fair and equitable sharing of the benefits of economic growth. But both global social policy frameworks often fail to acknowledge ecological limits that directly affect the limits of economic growth (Jackson & Webster, 2018).

Indeed, in a limitless world, traditional economic growth could go on forever, reducing poverty while at the same time not impeding the expectations of high-income groups that their wealth could also still grow. But there are clear limits to the use of the ecological resources that often generate wealth and economic growth. Indeed, six of nine planetary boundaries have been transgressed already (Richardson et al., 2023). Therefore, focusing global social policy discussions on limits might lead to the established assumptions and false win-win perspectives within social investment and inclusive growth being overcome (Schulze Waltrup, 2025b). Bob Deacon (2014), a pioneer in global social policy research in previous decades (Kaasch, 2019), emphasises the importance for the global social policy community of developing a resource consciousness. This highlights the need to consider limits, planetary boundaries and the global depletion of resources associated with the development of relatively comprehensive welfare state services in the Global North. Furthermore, he takes no-growth scenarios into consideration (without being very specific about them) and underlines how essential the redistribution not only of wealth but also of natural resources is for global social policy. Accordingly, global social policy has an important role to play in terms of redistributing (natural) resources and providing and coordinating social protection, as climate change will have a crucial impact on cross-border poverty. Generally, global social policy can mitigate the social pressure of climate change, which might be reflected in an increase in poverty and conflict and eventually in potential societal collapse. Therefore, global social policy can provide regulation, redistribution and basic rights to mitigate social pressures and distribute the costs and gains to enable a just and fair transition to climate change mitigation and adaptation (St. Clair & Lawson, 2014, p. 138).

Indeed, climate change has become an underlying issue in global social policy discourses, with IOs that, historically, did not prioritise climate change now addressing it as a major concern and linking it to their respective policy agendas. This is evident in various policy fields and measures, including migration, food and agricultural policy, childhood well-being and social cash transfers. As Silvana Lakeman (2021, p. 263) notes,

> IOs that previously had clearly assigned mandates, roles, and responsibilities are crossing (social) policy fields in response to the global threat of climate change. In this sense, global social policy is increasingly evolving into global eco-social policy, with actors forming interconnected governance structures that merge social policy with environmental policy agendas and prescriptions.

While Lakeman observes these developments on a rather general level, this book provides an assessment of two key global policy actors from an eco-social policy perspective while also revealing novel findings as part of four different analytical arguments (set out below) that contribute to both literature streams: that of global (social) policy studies and that of sustainable welfare and eco-social policy.

The following section outlines the structure of the book, explains how it assesses the OECD and World Bank from an eco-social policy perspective and presents its main analytical arguments and empirical findings. Chapter 2 provides an overview of the theoretical concepts that are applied at different levels of analysis and make sense of global policy-making in IOs on a broader scale, beyond those specific to eco-social or environmental and social policy. The chapter embeds research in world polity and discourse theory, a prerequisite for understanding this broader context, which is one of discursive world political struggles between different policy perspectives that try to institutionalise themselves as "taken-for-granted" within a world-societal context of perceiving what is "appropriate", and what is not. In other words, the book engages with world polity and discourse theory to study the extent to which global policy settings are being modified and converted, discursive world polity content is changing and eco-social policy ideas and models are being institutionalised as best practices at the global level by IOs. Moreover, applying discourse analysis, this book also highlights policy discourses that are systematically neglected and silenced by the OECD and World Bank within the process of environmental and social policy integration. This speaks to more radical policy discourses that question the above-mentioned conditions of the global political economy and conceive them as being responsible for the emergence of ecological and social crises. In order to illustrate the dynamic struggle of different and opposing policy perspectives within the two organisations' eco-social policy discourses, the book draws on the concept of the double movement developed by Karl Polanyi (2001) as a heuristic frame, besides making use of the concept of policy paradigms to assess the degree of change embedded within both organisations' eco-social policy discourses, asking whether there is a unified global eco-social policy paradigm.

Chapter 3 develops an eco-social policy typology that serves as a heuristic framework for comparing and assessing the policy discourses of the OECD and the World Bank from an eco-social perspective. The typology provides heuristic guidance on what types of policy discourses can be labelled as eco-social and to what extent these discourses contain eco-social policy potential and ideas. First, the chapter provides an overview of interconnections and tensions between environmental and social policy and then proceeds to review four different eco-social policy perspectives, presenting them as a typology that differentiates between them based on their relationship to public modes of governance and economic growth as well as their transformative potential. The discussion begins with the umbrella notion of "green growth", which encompasses two distinct eco-social policy perspectives: "green economy", which favours increasing the role of private actors and market-based solutions, and "green Keynesianism", which favours increasing the scope of government intervention. It then outlines "recomposing consumption and production" and "post-growth" as consecutive eco-social policy perspectives (Gough, 2017; Tienhaara, 2018). The chapter also discusses these different perspectives in relation to each other indicating how they represent "different degrees of eco-social policy" (Krause, 2021).

Chapters 2 and 3 indicate the different theories and concepts that this book applies at two different levels of the analysis. On a broader level covering global policy dynamics of policy-making, the theoretical framework (Chap. 2) is understood as delineating the social context and the conditions in which the two IOs' discursive activities unfold. This means that, in particular, the combinations of world polity theory, discourse theory, the concept of policy paradigm and the dynamic of the double movement provide the general framing in which the emergence of new settings of new global policy discourses takes place. This context is one of a discursive struggle between partly antagonistic policy perspectives at the global level. On another somewhat more practical and policy-oriented level, the eco-social policy typology developed in Chap. 3 delineates the actual focus of the empirical analysis. The creation of this typology is based on an extensive literature review of different perspectives and concepts from environmental economics, political economy, policy studies and ecological economics. The book applies this typology as a basis of comparison in order to study the eco-social policy content of the selected documents from the organisations.

The sampling of documents was primarily based on eco-social policy knowledge derived from the typology developed in Chap. 3. An initial pool of documents was collected through a search of the electronic libraries of both organisations. Document collection was stopped when no more new insights emerged, indicating that theoretical saturation had been reached (Glaser & Strauss, 1967). Consequently, further sampling of documents for analysis depended on the potential for new insights or variations on previously gathered information. This means that a targeted selection process was used to compile a dataset of documents, which primarily covers the period from 2011 to 2021, with a few selected documents from before 2011. Background conversations with individuals who worked on the documents were also conducted to support the analysis.

While the analysis is conceptually grounded in discourse theory, the documents are analysed using qualitative content analysis as the primary method with a particular focus on the techniques of structuring, summarising and classification (Mayring, 2015). Consequently, the data material is structured according to certain conceptual criteria for which the eco-social policy typology serves as the heuristic frame. This creates summarising descriptions based on a deductive application of the typology while at the same time ensuring openness and receptiveness with regard to how the organisations interpret the concepts that are part of the typology. The goal, at the methodological level, is to create text corpora that are comprehensible but still a good representation of the original material. This also distils various aspects within the empirical material based on the typology. The text corpora based on the qualitative content analysis are presented in Chaps. 4 and 5. They provide an empirical reconstruction of the OECD's and World Bank's eco-social policy and review the concepts that the organisations prescribe in global environmental and social policy. Applying the eco-social policy typology means comparing and contrasting the organisations' policy discourse with the different eco-social policy perspectives that are part of the typology. Accordingly, the chapters on the organisations' eco-social policy discourses explore their respective understandings—green growth, green economy, green Keynesianism, recomposing consumption and production and post-growth—while interconnections and tensions between environmental and social policy are discussed as part of their understanding of green Keynesianism. This means that the chapters provide mainly empirical, in-depth reconstructions of different policy notions that are considered crucial for understanding the degree and categories of eco-social policy advocated by the OECD and the World Bank. Chapter 6 explicitly compares the organisations' eco-social policy discourses. The comparison highlights that there are crucial differences concerning particular themes that result from the different policy foci of these organisations. However, in general, there are more similarities than differences between the IOs' discourses.

Chapter 7 sets out the empirical findings and categorisations of both organisations' policy discourses through four analytical arguments. First, that the eco-social policy discourses prescribed by the two organisations are not coherent sets of ideas but instead propagate different sets of ideas characterised by the dynamic of Polanyi's double movement. The first movement consists of "the principle of economic liberalism" (Polanyi, 2001, p. 138) permeated by a laissez-faire approach that largely favours the interests of the trading class, while the second, also termed the countermovement, is driven by "the principle of social protection" (ibid.) that largely favours protective legislation and government intervention. Drawing on other theoretical and historical research that transfers this thinking to the contemporary global governance system, this book provides empirical evidence for such an analytical perspective within contemporary global governance and argues that the double movement is present within both organisations' eco-social policy discourses, in other words, is reflected in the OECD's and World Bank's conception of green economy and green Keynesianism, respectively, which are both integral parts of each organisation's eco-social policy perspective. Accordingly, the book argues that green economy resonates with the first movement by establishing a self-regulating (green) market and increasing the scope of private and market-based solutions while curtailing the scope of public policy to only providing a framework in which the self-regulating (green) market can flourish. Green Keynesianism, on the other hand, resonates with the countermovement by calling for comprehensive government intervention and a profound set of social protection systems. In other words, the chapter argues that the double movement can be found "under the same roof" (Mendly, 2020, p. 515) in each of the two IOs. The dynamic interplay between the two movements is a crucial mechanism for stabilising the existing capitalist mode of production within but also beyond those organisations, as the second and third analytical arguments suggest.

Second, Chap. 7 assesses the OECD's and World Bank's eco-social policy discourses in terms of their potential to drive a social-ecological transformation. Both organisations still advocate substantial elements of "liberal globalism", which is largely reflected in the first movement and their green economy understanding. However, as part of their green Keynesian understanding, or of the countermovement within the double movement, they reform their notion of liberal globalism, favouring an entrepreneurial state with comprehensive government intervention. Such attempts at reform signal a move towards the policy agenda of "green embedded liberalism", which rejects both right- and left-wing deglobalisation attempts and conceives the authority of public agencies as crucial to actively steering transnational capital and labour developments towards ecological protection. However, both organisations fall short of driving a more radical social-ecological transformation which would include attempts to decommodify nature and labour to invoke post-growth policies that follow criteria of socio-economic democratisation, emphasising community practices, selective economic deglobalisation and a cooperative global governance logic (Brand et al., 2020; Foundational Economy Collective, 2018; Novy, 2022).

Third, Chap. 7 contends that both organisations' eco-social policy discourses conform to what Brand et al. (2020) have called a "new critical orthodoxy". Accordingly, the chapter argues that the existing eco-social orthodoxy acknowledges that there are severe and extensive problems and crises, such as inequality and ecological collapse that require profound social changes. However, it then proposes solutions that rely on existing institutions and capitalist dynamics, assuming that these can solve contemporary problems and crises if adapted in specific ways. Accordingly, the chapter argues that green Keynesianism represents such an adaptation and shift towards a "new" capitalist mode of production in which the state reasserts control over the market to promote the logics of capital accumulation and logics of valorising nature. This shift could signal a shift towards a synergetic relationship between the two movements that results in the stabilisation of the "organisation of production that the market had called into being" (Polanyi, 2001, p. 136)—in other words, the capitalist mode of production in global governance. This discursive strategy also silences more radical policy approaches, such as post-growth.

Fourth, the chapter assesses the organisations' eco-social policy discourses in terms of policy paradigm change (Buch-Hansen & Carstensen, 2021; Hall, 1993). Both organisations' advocacy of green Keynesianism and the struggle between green economy and green Keynesianism within the double movement dynamic may indicate a third-order paradigm change or at least a conflict between third-order paradigms vying for dominance. A third-order paradigm change is a profound paradigm shift in guiding institutions and ideas within policy-making, whereas a second-order change involves changes in policy instruments, and a first-order one, a change in policy settings. While third-order paradigm change occurs within a capitalist logic, as illustrated by the change between Keynesianism and monetarism studied by Hall (1993), it falls short of a fourth-order change of "general systemic logics" (Buch-Hansen & Carstensen, 2021, p. 312), which would abandon the economic growth-based capitalist mode of production and move towards post-growth policy ambitions to radically decommodify nature and labour. This book suggests that paradigm change is not necessarily an abrupt process but rather a long-term struggle between paradigms that can coexist and relate to each other within one organisation or, more broadly, in policy fields and contexts. However, to account for all eco-social policy ideas and notions, not just green economy and green Keynesianism, it cannot be said that there is a coherent and unified "global eco-social policy paradigm". Consequently, the book proposes to perceive the eco-social policy constellation of ideas promoted by both organisations as a "web of loose [eco-social] ideas" (Von Gliszczynski, 2015, p. 67ff.) in global governance that is shaped by the double movement dynamic.

References

Béland, D., & Orenstein, M. A. (2013). International organizations as policy actors: An ideational approach. *Global Social Policy: An Interdisciplinary Journal of Public Policy and Social Development, 13*(2), 125–143. https://doi.org/10.1177/1468018113484608

Bohnenberger, K. (2020). Money, vouchers, public infrastructures? A framework for sustainable welfare benefits. *Sustainability, 12*(2). https://doi.org/10.3390/su12020596

Brand, U., Görg, C., & Wissen, M. (2020). Overcoming neoliberal globalization: Social-ecological transformation from a Polanyian perspective and beyond. *Globalizations, 17*(1), 161–176. https://doi.org/10.1080/14747731.2019.1644708

Bridgen, P. (2023). An eco-social solution to energy poverty? Substance and symbolism in the England's use of domestic energy efficiency policy to achieve social and environmental synergies, 1997–2023. *European Journal of Social Security, 25*(4), 388–407. https://doi.org/10.1177/13882627231195722

Buch-Hansen, H., & Carstensen, M. B. (2021). Paradigms and the political economy of ecopolitical projects: Green growth and degrowth compared. *Competition & Change, 25*(3–4), 308–327. https://doi.org/10.1177/1024529420987528

Budowski, M., & Künzler, D. (2022). Editorial: Social sustainability: An ambiguous concept, areas of tension and ways forward. *sozialpolitik.ch, 2022*(1), 1–19. https://doi.org/10.18753/2297-8224-185

Cigna, L., Fischer, T., Abuannab, E. H., Heins, E., & Rathgeb, P. (2023). Varieties of just transition? Eco-social policy approaches at the international level. *Social Policy and Society, 22*(4), 730–746. https://doi.org/10.1017/S1474746423000192

Cotta, B. (2024). Unpacking the eco-social perspective in European policy, politics, and polity dimensions. *European Political Science, 23*(1), 1–13. https://doi.org/10.1057/s41304-023-00453-6

De Francesco, F. (2014). Transfer agents, knowledge authority, and indices of regulatory quality: A comparative analysis of the World Bank and the organisation for economic co-operation and development. *Journal of Comparative Policy Analysis: Research and Practice, 18*(4), 350–365. https://doi.org/10.1080/13876988.2014.882648

De Francesco, F., & Guaschino, E. (2019). Reframing knowledge: A comparison of OECD and World Bank discourse on public governance reform. *Policy and Society, 39*(1), 113–128. https://doi.org/10.1080/14494035.2019.1609391

Deacon, B. (2007). *Global social policy & governance.* Sage.

Deacon, B. (2014). Toward a transformative global social policy? In A. Kaasch & P. Stubbs (Eds.), *Transformations in global and regional social policies* (pp. 201–217). Palgrave Macmillan.

Deeming, C. (2021). The struggle for social sustainability. In C. Deeming (Ed.), *The struggle for social sustainability. Moral conflicts in global social policy* (pp. 291–345). Policy Press.

Deeming, C., & Smyth, P. E. (Eds.). (2018). *Reframing global social policy. Social investment for sustainable and inclusive growth.* Policy Press.

Di Leva, C. E. (2021). The challenge and promise at the intersection of environmental and social policies: How the World Bank established a policy framework that fully integrates environmental and social concerns. *Global Social Policy, 21*(2), 344–348. https://doi.org/10.1177/14680181211019170

Fanning, A. L., O'Neill, D. W., & Büchs, M. (2020). Provisioning systems for a good life within planetary boundaries. *Global Environmental Change, 64.* https://doi.org/10.1016/j.gloenvcha.2020.102135

Fanning, A. L., O'Neill, D. W., Hickel, J., & Roux, N. (2022). The social shortfall and ecological overshoot of nations. *Nature Sustainability, 5*(1), 26–36. https://doi.org/10.1038/s41893-021-00799-z

Foundational Economy Collective. (2018). *Foundational economy. The infrastructure of everyday life.* Manchester University Press.

Glaser, B., & Strauss, A. L. (1967). *The discovery of grounded theory: Strategies for qualitative research.* Aldine de Gruyter.

Gough, I. (2013). Climate change, social policy, and global governance. *Journal of International and Comparative Social Policy, 29*(3), 185–203. https://doi.org/10.1080/21699763.2013.852128

Gough, I. (2017). *Heat, greed and human need. Climate change, capitalism and sustainable wellbeing.* Edward Elgar Publishing.

Greer, S., Elliott, H., & Oliver, R. (2020). Differences that matter: Overcoming methodological nationalism in comparative social policy research. In *Theory and methods in comparative policy analysis studies* (pp. 267–288). Routledge.

Hall, P. A. (1993). Policy paradigms, social learning, and the state: The case of economic policymaking in Britain. *Comparative Politics, 25*(3), 275–296. https://doi.org/10.2307/422246

Hammer, P. J. (2013). *Change and continuity at the World Bank: Reforming paradoxes of economic development.* Edward Elgar.

Hirvilammi, T., & Koch, M. (2020). Sustainable welfare beyond growth. *Sustainability, 12*(5). https://doi.org/10.3390/su12051824

Hirvilammi, T., Peltomaa, J., Pihlajamaa, M., & Tiilikainen, S. (2023). Towards an eco-welfare state: Enabling factors for transformative eco-social initiatives. *European Journal of Social Security, 25*(4), 445–463. https://doi.org/10.1177/13882627231195724

Hynes, W. (2021). A systemic socio-ecological recovery from Covid-19. *Global Social Policy, 21*(2), 335–338. https://doi.org/10.1177/14680181211019166

Jackson, T., & Webster, R. (2018). Limits to growth revisited. In C. Deeming & P. Smyth (Eds.), *Reframing global social policy. Social investment for sustainable and inclusive growth* (pp. 295–322). Policy Press.

Kaasch, A. (2019). Introduction to a special issue in memory of Bob Deacon: Where do we stand in global social policy studies? *Global Social Policy, 19*(1–2), 4–20. https://doi.org/10.1177/1468018119847336
Kaasch, A., & Martens, K. E. (2015). *Actors and agency in global social governance*. Oxford University Press.
Kaasch, A., & Schulze Waltrup, R. (2021). Introduction: Global eco-social policy: Contestation within an emerging policy era? *Global Social Policy, 21*(2), 319–322. https://doi.org/10.1177/14680181211019152
Koch, M. (2022). Social policy without growth: Moving towards sustainable welfare states. *Social Policy and Society, 21*(3), 447–459. https://doi.org/10.1017/S1474746421000361
Köhler, G. (2016). Assessing the SDGs from the standpoint of eco-social policy: Using the SDGs subversively. *Journal of International and Comparative Social Policy, 32*(2), 149–164. https://doi.org/10.1080/21699763.2016.1198715
Krause, D. (2021). Uncovering different degrees of eco-social policy. *Global Social Policy, 21*(2), 332–334. https://doi.org/10.1177/14680181211019165
Lakeman, S. (2021). IOs and climate change: Toward global eco-social policy. In K. Martens, D. Niemann, & A. Kaasch (Eds.), *International organizations in global social governance* (pp. 255–273). Palgrave Macmillan.
Lamb, W. F., Antal, M., Bohnenberger, K., Brand-Correa, L. I., Müller-Hansen, F., Jakob, M., et al. (2020). What are the social outcomes of climate policies? A systematic map and review of the ex-post literature. *Environmental Research Letters, 15*(11). https://doi.org/10.1088/1748-9326/abc11f
Laruffa, F., & Nullmeier, F. (2025). A model to follow? The EU and global eco-social policy. *Global Social Policy, 25*(1), 151–169. https://doi.org/10.1177/14680181241246935
Mahon, R., & McBride, S. (2009). Standardizing and disseminating knowledge: The role of the OECD in global governance. *European Political Science Review, 1*(1), 83–101. https://doi.org/10.1017/S1755773909000058
Mandelli, M. (2022). Understanding eco-social policies: A proposed definition and typology. *Transfer: European Review of Labour and Research, 28*(3), 333–348. https://doi.org/10.1177/10242589221125083
Mandelli, M. (2025). The comparative politics of just transition policies: Building green–red winning coalitions in Spain and Ireland. *Journal of European Public Policy*, 1–26. https://doi.org/10.1080/13501763.2024.2446364
Markkanen, S., & Anger-Kraavi, A. (2019). Social impacts of climate change mitigation policies and their implications for inequality. *Climate Policy, 19*(7), 827–844. https://doi.org/10.1080/14693062.2019.1596873
Martens, K., Niemann, D., & Kaasch, A. (Eds.). (2021). *International organizations in global social governance*. Palgrave Macmillan.
Mayring, P. (2015). *Qualitative Inhaltsanalyse* [Qualitative content analysis] (12th ed.). Beltz.

Mendly, D. (2020). Global governance and the double movement. *Global Governance: A Review of Multilateralism and International Organizations, 26*(3), 500–521. https://doi.org/10.1163/19426720-02603006

Novy, A. (2022). The political trilemma of contemporary social-ecological transformation—Lessons from Karl Polanyi's *The Great Transformation*. *Globalizations, 19*(1), 59–80. https://doi.org/10.1080/14747731.2020.1850073

O'Neill, D. W., Fanning, A. L., Lamb, W. F., & Steinberger, J. K. (2018). A good life for all within planetary boundaries. *Nature Sustainability, 1*(2), 88–95. https://doi.org/10.1038/s41893-018-0021-4

Orenstein, M. A. (2008). *Privatizing pensions*. Princeton University Press.

Polanyi, K. (2001 [1944]). *The Great Transformation. The political and economic origins of our time*. Beacon Press.

Richardson, K., Steffen, W., Lucht, W., Bendtsen, J., Cornell, S. E., Donges, J. F., et al. (2023). Earth beyond six of nine planetary boundaries. *Science. Advances, 9*(37). https://doi.org/10.1126/sciadv.adh2458

Sabato, S., & Mandelli, M. (2024). Towards an EU framework for a just transition: Welfare policies and politics for the socio-ecological transition. *European Political Science, 23*(1), 14–26. https://doi.org/10.1057/s41304-023-00458-1

Schmelzer, M. (2016). *The hegemony of growth: The OECD and the making of growth*. Cambridge University Press.

Schulze Waltrup, R. (2025a). An eco-social policy typology: From system reproduction to transformation. *Global Social Policy, 25*(1), 17–35. https://doi.org/10.1177/14680181231205777

Schulze Waltrup, R. (2025b). Questioning global social policy leitmotifs: Europe as a post-growth reference point amid planetary boundaries transgression. *Global Social Policy, 25*(1), 176–179. https://doi.org/10.1177/14680181241311994

Schulze Waltrup, R., Moore, M., & Paulsen, T. (2023). Eco-social policy in the global political economy: Analysing shifting discourses on agricultural subsidies. *European Journal of Social Security, 25*(4), 506–522. https://doi.org/10.1177/13882627231206002

Schulze Waltrup, R., Gengnagel, V., Zimmermann, K., & Kaasch, A. (2025). Setting the scene in Europe: Emerging national and transnational social policy responses to climate change. *Global Social Policy, 25*(1), 3–16. https://doi.org/10.1177/14680181241312557

Smyth, P., & Deeming, C. (2018). Conclusion: Towards a new global social policy framework? In C. Deeming & P. Smyth (Eds.), *Reframing global social policy. Social investment for sustainable and inclusive growth* (pp. 323–342). Policy Press.

St. Clair, A. L., & Lawson, V. (2014). Poverty and climate change: The three tasks of transformative global social policy. In A. Kaasch & P. Stubbs (Eds.), *Transformations in global and regional social policies* (pp. 134–152). Palgrave Macmillan.

Stone, D. (2003). The "Knowledge Bank" and the Global Development Network. *Global Governance: A Review of Multilateralism and International Organizations, 9*(1), 43–61.

Tienhaara, K. (2018). *Green Keynesianism and the global financial crisis.* Routledge.

Von Gliszczynski, M. (2015). *Cash transfers and basic social protection.* Palgrave Macmillan.

Wade, R. H. (2002). US hegemony and the World Bank: The fight over people and ideas. *Review of International Political Economy, 9*(2), 215–243. https://doi.org/10.1080/09692290110126092

World Commission on Environment and Development. (1987). *Our Common Future.* Oxford University Press.

Zimmermann, K., & Graziano, P. (2020). Mapping different worlds of eco-welfare states. *Sustainability, 12*(5). https://doi.org/10.3390/su12051819

CHAPTER 2

A Theoretical Framework of Global Policy Dynamics

This book employs different theoretical and conceptual frameworks at various levels of analysis. While the developed eco-social policy typology (Chap. 3) helps to assess the OECD and World Bank from an eco-social policy perspective in terms of the content and discourse that both organisations promote, the theoretical framework of global policy dynamics that the current chapter develops helps to make sense of the construction processes underlying this eco-social policy content. Since it reconstructs certain perceptions of the interconnections between policy fields, the book creates a theoretical framework that goes beyond the specific notions of eco-social, environmental and social policy. This framework transcends the dynamics of these individual policy fields. It helps to understand the extent to which the eco-social policy notions in question represent a change from existing global policy notions and to explain the underlying dynamics responsible for creating conflictual eco-social policy perspectives within organisations.

On a broader theoretical level, this study is influenced by world polity and discourse theory. This contextualises the empirical analysis against a backdrop of world political struggles between different policy perspectives that aim to institutionalise themselves as universally valid thus rendering themselves self-evident within the perception of what appropriate policy-making is and what it is not. Accordingly, this study draws on discourse theory to reconstruct the kind of policy perspectives that are structurally

R. Schulze Waltrup, *Eco-Social Policy in the OECD and the World Bank*, https://doi.org/10.1007/978-3-031-95429-0_2

neglected within the creation of eco-social policy at the global level of the two IOs. In order to further interpret and illustrate the dynamic and discursive struggle between the different and opposing perspectives within both organisations' eco-social policy discourses, this study draws on Polanyi's (2001) concept of the double movement. The double movement concept has a strong explanatory value concerning the dynamic of opposing policy perspectives within both organisations (see Sect. 7.1). In order to assess the degree of change that takes place within the creation of an eco-social policy perspective in both organisations, this book draws on the concept of different orders of policy paradigm change. This is also connected to the discursive struggle between opposing policy paradigms and the double movement dynamic (Sect. 7.4).

2.1 World Polity

In order to understand the world-societal context in which the construction of global eco-social policy models takes place, this section provides a brief review of world polity theory. The world polity is a rationalised global cultural order that consists of certain policy models that are perceived as self-evident and translated to different actors worldwide, from national governments and transnational organisations to local individual actors. Unlike most theories of the international system that regard nation-states as units consisting of and constituted by the individual actors living within them who act according to a certain logic of cooperation and competition, world polity theory assumes that the system of nation-states is instead constituted by a rationalised global cultural order (Meyer et al., 1997). The assumption is that national governments and other actors seek to legitimise themselves on the basis of these institutionalised global models that are part of the world polity, such as citizenship, socio-economic development and rationalised justice (with potentially varied interpretations of what those concepts actually entail). The world polity provides a relatively small and specific spectrum of legitimate behaviours, and certain policy practices are institutionalised in the sense of being part of a process in which actions and thinking patterns are granted normative and cognitive validity and accepted as common sense and self-evident (Meyer et al., 1987, p. 13). Due to the serious political implications of this process, the development of such a culture is often characterised by profound contestation and conflicts because actors want to impose different versions and interests and they try to inject them into these global rationalised models in order to present them as universal and self-evident (Lechner & Boli, 2005, p. 24). IOs such as the OECD and the World Bank play a crucial role in creating and disseminating these taken-for-granted and self-evident policy models. Based on their authority, knowledge and expertise, they present themselves as "disinterested rationalized others" (Meyer et al., 1997, p. 165) in order to be able to provide fairly universal policy prescriptions that entail the idea of serving the general interest of world society. Applying this perspective, this book reconstructs eco-social policy models and examines the extent to which they are institutionalised at the global level, acknowledging the OECD and World Bank as crucial actors driving the formation and creation of these global policy models.

2.2 The Discursive Construction of Global Policy Models

A crucial theoretical complement to the study of construction processes of global policy models is discourse theory (Meyer, 2009, p. 43). Discourse centres on language as being essential in constructing and institutionalising a particular social reality. While studying discourses within a world polity framework, it is essential to recognise that not all discourses are institutionalised within the world polity, but any world polity framework consists of discourses. While many discourse elements do not have the same impact as institutionalised ones, they may attempt to contest, shift and change them (Phillips et al., 2004, p. 638). Combining discourse theory and world polity theory is particularly valuable for gaining a deeper understanding of the construction of taken-for-granted practices that are deemed universally valid on the global level. Accordingly, this combination aims to analyse how global policy models are reconstituted, reproduced or changed within processes of knowledge generation. In that process, it is "historically specific rules—embedded in language, practices, and objects" (Tag, 2013, p. 33)—that render the appearance of specific expressions likely and, accordingly, have the potential to modify and transform global policy settings. With such a combination, the book aims to study specific discursive and communicative events (micro events) in order to draw conclusions about the actual construction of global policy models and how these cultural elements are performed as universally applicable (macro events).

Discourse theory perspectives adopt a "conflictual picture" that acknowledges different discourses that are in conflict with each other and coexist in relation to each other despite antagonistic attempts to define truth and the rules of the game (Jørgensen & Philips, 2002, p. 13). In particular, critical discourse analysis (CDA) aims to explain discursive (non-)change and reconstructs the often naturalised and taken-for-granted political strategies to which a discourse belongs (Fairclough, 2010, p. 30f.), which is also a crucial objective in this book. This is underpinned by the dialectical assumption that micro-events, such as communicative events, are both constituted by and constitutive of overarching structures. Accordingly, micro-events are always part of, or refer to, an overarching discourse, paradigm or ideology (Fairclough, 2010, p. 31). Successful ideological practices superficially dissociate themselves from the groups that benefit from them and are perceived as "societal consensus". They are naturalised, depoliticised and are no longer recognised as practices that solely benefit certain groups but are often framed as being in the general interest of society. However, this discursively obscures the fact that these policy models often directly benefit particular groups while disadvantaging others (see Sects. 7.2 and 7.3 on how eco-social policy models within the OECD and World Bank reaffirm a capitalist mode of production to the detriment of other perspectives).

CDA is also closely related to the concept of governmentality (Fairclough, 2010; Foucault, 1991). While the concept of governmentality is not applied in this book with the aim of providing an overall interpretation of the dynamics of both organisations' eco-social policy discourses, Section 5.3.3 applies the concept to interpreting the governing mechanisms of one of the World Bank's crucial documents, the Environmental and Social Framework. CDA and the concept of governmentality highlight how certain technologies and social science methodologies can be used in discourses to create subjects and support their autonomy, while managing the conduct of these entities and, in doing so, steering them towards certain practices without breaching their formal autonomy. There are various mechanisms that contribute to this process, such as methods of assessing, monitoring, reporting, creating indicators and measuring transparent standardised information. These are more broadly considered to be governmental technologies as measures and programmes of calculations and procedures (Fairclough, 2010; Miller & Rose, 2008; Turnhout et al., 2014). This reflects an indirect and diffuse disposition of various techniques and tactics to steer the development of the targeted actor(s). In general, and in the context of IOs, governmentality particularly emphasises the important role of the language and technical means available to IOs in directing the conduct of nation-states (Merlingen, 2003, p. 371). Section 5.3.3 will briefly expand on these considerations before providing an interpretation of the World Bank's Environmental and Social Framework using the lens of governmentality.

The dialectical and potentially conflictual relationship between micro-events and overarching structures, mentioned above, is key to CDA and remarkably complementary to Polanyi's concept of the double movement.

2.3 The Double Movement: Making Sense of Opposing Policy Perspectives

In order to further illustrate and theorise the dynamic and discursive struggle between opposing perspectives within the eco-social policy discourses of the OECD and World Bank, this book draws on the concept of the double movement (Polanyi, 2001), which holds crucial explanatory value for conceptualising the opposing policy dynamics within the discourses of the organisations. In Sect. 7.1, this book argues that the eco-social policy discourses of both organisations entail different perspectives on the double movement and stabilise a capitalist mode of production through a policy constellation of interventions and regulations.

While Polanyi describes the concept of a double movement as part of his historical reconstruction of the emergence of a self-regulating market during the industrialisation of England in the nineteenth and twentieth centuries, other scholars have proven its validity and relevance in understanding contemporary politics as well, in both the Global South and North (e.g. Atzmüller et al., 2019; Blyth, 2002; Burawoy, 2015; Goodwin, 2018). Polanyi proposes to make sense of policy-making based on the concept of the double movement. The first movement consists of "the principle of economic liberalism" (Polanyi, 2001, p. 138) permeated by an economic laissez-faire approach that largely favoured the interests of the trading class and was based on the propagation of free trade, while the second, also termed the countermovement, is driven by "the principle of social protection" (Polanyi, 2001, p. 138) that largely favoured the interests of (rural) workers and policy interventions, such as protective legislation. While introducing the notion of the double movement, Polanyi reconstructs the emergence and establishment of the idea and practice of a self-regulating market as part of the first movement during the industrial revolution in England and highlights the detrimental social and environmental consequences that these specific market operations created. As Polanyi argues, these detrimental consequences have led to the emergence of a protective countermovement that aims to re-embed the market into society, preventing the "running of society as an adjunct to the market" (Polanyi, 2001, p. XXIV) with its consequent destructive effects.

The general idea of a self-regulating market is that there should be no policy action that would in any way restrict or steer market operations. Accordingly, there must be no regulation of price, supply or demand while regulation's main task is to create "conditions which make the market the only organising power in the economic sphere" (Polanyi, 2001, p. 72). However, Polanyi shows that the self-regulating market ends up "annihilating the human and natural substance of society" (Polanyi, 2001, p. 3). The reason for this is that the self-regulating market forces elements, such as labour, land (nature), and money to be treated as regular commodities. Usually, regular commodities are goods that are produced in order to be exchanged on the market. However, labour, land and money cannot be adequately characterised as being produced solely for exchange on the market: they are simply too important to essential aspects of human reproduction and well-being to be simply treated as being exchangeable on the market like regular commodities. Polanyi (2001, p. 71ff.), therefore, conceptualises them as fictitious commodities. According to him, it is precisely the subjugation of fictitious commodities to the self-regulating market that has led to severe social and environmental pressures. He claims that, if the self-regulating market remains unchecked and is not opposed by social protection, its unconstrained operations will lead to "the demolition of society" (Polanyi, 2001, p. 76):

> Undoubtedly, labor, land, and money markets *are* essential to a market economy. But no society could stand the effects of such a system of crude fictions even for the shortest stretch of time unless its human and natural substance as well as its business organization was protected against the ravages of this satanic mill. (Polanyi, 2001, p. 76f.)

Hence, as a mechanism of the "self-protection of society" (Polanyi, 2001, p. 135), the countermovement operates as "more than the usual defensive behavior of a society faced with change" (Polanyi, 2001, p. 136) as it is a protective countermovement that not only saves society but also, within a dialectic and dynamic interplay between both movements, stabilises the "organization of production that the market had called into being" (Polanyi, 2001, p. 136). In doing so, it also stabilises a capitalist mode of production through being characterised by a comprehensive policy constellation of interventions and regulations. Such a countermovement, for example, led to the establishment of the "embedded liberalism" (Ruggie, 1982) enshrined in the post-war order of comprehensive welfare state services and programmes in the Global North (Atzmüller et al., 2019; Deutschmann, 2019), which Heikki Patomäki (2014), for example, sees as a synthesis of the dialectics of the double movement. Referring to this, the book later argues that both IOs' eco-social policy discourses reflect the dynamic of the double movement and thereby stabilise and adapt the contemporary capitalist mode of production within global governance. This is largely reflected by a move towards green Keynesianism (Chap. 7).

Some scholars have explicitly applied such a Polanyian perspective to the global governance system (Patomäki, 2014; Mendly, 2020). Dorttya Mendly (2020) claims that global governance has become a hegemonic discourse precisely because it integrates the dynamic of the countermovement within its structures and its policy discourses. Accordingly, there is a crucial balance within global governance between establishing and reproducing global market-based policies and a social protection approach that aims to protect society from the severe consequences of an unleashed market. Furthermore, there are a variety of actors within global governance to whom the different perspectives of the double movement can be attributed (Mendly, 2020). However, within global governance, market-based types of governance play a more dominant role in policy discourses, and, as a general rule, interventions and regulations that try to impede the operation of a self-regulating market are more likely to be reactive and merely respond to this more dominant position. This means that, generally, global governance also integrates the concerns of, for example, social movements whose object is to avoid environmental degradation and potential social stress—though this is often done to reproduce, reinforce and adapt the existing capitalist mode of production (Mendly, 2020).

The concept of the double movement offers a framework for comprehending the often heterogeneous and partially conflicting policy discourses in the global governance system (Mendly, 2020; Patomäki, 2014). While Mendly (2020) provides a general theoretical and historical analysis of the double movement, she calls for more empirically grounded research on the topic. She briefly notes, without going into much detail, that within the broad United Nations (UN) system, the two motivations for the double movement coexist, often "under the same roof" (Mendly, 2020, p. 515). This book contributes to this perspective by examining how those motivations are reflected in the eco-social policy discourses of the OECD and World Bank, respectively (see Sect. 7.1).

2.4 Policy Paradigms

While the concept of the double movement is crucial to theorising the dynamics of the opposing policy perspectives of the IOs' eco-social policy discourses, this book also draws on the concept of policy paradigm change to assess the degree of change in the organisations' eco-social policy debates. Section 7.4 investigates more specifically whether the eco-social policy discourses identified can be considered as a unified global eco-social policy paradigm.

When discussing social learning, Peter Hall (1993) introduces the concept of policy paradigms. He states that policy-making has three dimensions. First, there is an overarching goal guiding the general policy-making process that represents the third order of policy-making. Second, there are techniques or instruments for achieving the goal, that represent the second order. Third, there are "precise settings of these instruments" (Hall, 1993, p. 278), that represent the first. While changes within the latter two dimensions (the first and second orders) are quite frequent and represent the usual process of policy-making, changes in the former dimension (the third order) are quite rare and represent bigger shifts in the goals and concepts guiding the policy-making process. This overarching goal is a policy paradigm that structures the way in which policy-makers perceive policy problems, their solutions and how and in what words they communicate about them: "Like a *Gestalt*, this framework is embedded in the very terminology through which policymakers communicate about their work, and it is influential precisely because so much of it is taken for granted and unalienable to scrutiny as a whole" (Hall, 1993, p. 279). In that sense, facts or evidence potentially undermining the policy paradigm are often overlooked, ignored or reinterpreted to make them congruent with the existing paradigm (Princen & Van Esch, 2015, p. 357f.).

In Hall's original conception, a change of policy paradigms is to be considered as a third-order change. Changes in the use of techniques and instruments and in the "precise settings of these instruments" (Hall, 1993,

p. 278) usually reflect a simple retention and reproduction of established policy patterns, while a change of paradigms reflects "a more disjunctive process associated with periodic discontinuities in policy" (ibid., p. 279). However, there is no consensus on how paradigm change takes place. There are two lines of research. One (conforming to Hall's original conception) suggests that paradigmatic change is a rather abrupt process; the other, that paradigm change takes place fairly gradually (Carson et al., 2009; Princen & Van Esch, 2015; Skogstad & Schmidt, 2017): "Far from making a clean slate of the past, a new societal paradigm must in effect be composed of previous cognitive and normative structures" (Surel, 2000, p. 508). In a similar sense, paradigm change via layering suggests that changes within the paradigm come in addition to already existing concepts of the paradigm (Skogstad & Schmidt, 2017, p. 12). From this perspective, Diercks et al. (2019) reconstruct the development of innovation policy paradigms as a layering process. They argue that a new transformative innovation policy paradigm has been established in policy debates, but this new policy paradigm does not replace earlier ones but is rather layered upon traditional innovation and technology policy paradigms, so that earlier examples are not fully replaced. As this book argues in Sect. 7.4, the empirical findings of this book suggest that policy paradigm change takes place fairly gradually and is shaped by a dynamic and conflictual struggle between antagonistic paradigms within the OECD and World Bank.

Nor is there any research consensus on the incommensurability and uniqueness of paradigms. Hall's original conception suggests that there is hardly any chance of paradigms coexisting since they entail incommensurable assumptions which can hardly be remedied. However, as other research suggests and this book also reconstructs in the different subsections of Chap. 7, the empirical evidence indicates that mutually opposing paradigms can exist simultaneously, with one of them still being the dominant one (see also Princen & Van Esch, 2015). Sabatier and Jenkins-Smith (1993) argue that dominant paradigms or, to use their words, dominant belief systems compete with minority ones. Minority belief systems constantly try to challenge and contest the dominant system by referring to various cognitive concepts and normative ideas (Skogstad & Schmidt, 2017, p. 9).[1] Princen and Van Esch (2015, p. 371), for example, come to

[1] See also the work of Streeck and Thelen (2005) on incremental but potentially transformative change due to layering, conversation and drift, as well as their concept of displacement in regard to relatively sudden shifts.

the conclusion that the EU's Stability and Growth Pact is a mixture of policy paradigms. It is characterised by "varying dominance of one paradigm within a mix of paradigms". This suggests that (contrary to Hall's expectation), it is indeed possible that there are mixes of various paradigms. The empirical evidence of this book confirms this. As argued later, the coexistence of opposing paradigms within the OECD and World Bank is characterised by the dynamic of the double movement (Chap. 7).

Within studies of policy paradigms, the concept of discourse is crucial to understanding how they are constituted. Policy paradigms entail various discourses that serve the overall paradigmatic policy framework. In that sense, particular discourses express the institutionalised form of a policy paradigm. These paradigm-infused discourses enact a shared framework that provides sense-making as to how the world is constituted and what kind of policies are appropriate to solving perceived problems (Carson et al., 2009, p. 17). However, other discourses might emerge as well to change and tackle established discourses and policy paradigms (ibid.). Thus, studies examining paradigms also investigate to what extent other discourses can break the paradigmatic "intellectual path dependency in policy-making" (Blyth, 2001, p. 4) to institutionalise different sets of ideas and discourses (Skogstad & Schmidt, 2017, p. 3). A particularly important role is played by knowledge elites, who attempt to construct alternative paradigms on which those in power can base their actions. In this process, IOs have a specific role, that of setting "international knowledge standards" (Skogstad & Schmidt, 2017, p. 17). Moreover, IOs often have moral, legal and expert authority which enables them to construct and diffuse policy paradigms into domestic conditions. Accordingly, IOs can be crucial actors that use discourses and framing strategies either to question an existing paradigm or to support one when paradigms are being contested.

In Hall's and all the following work on paradigm change, the broadest and most consequential level of change is usually of the third-order variety.[2] However, this is still conceptualised as being change within a capitalist framework—from Keynesianism to monetarism—both of which are driven by economic growth. In order to conceptualise policy paradigm

[2] See also the literature on the advocacy coalition framework, which shares a similar concept of most consequential change to a certain degree. However, for a discussion of the similarities and differences between the study of policy paradigms and the advocacy coalition framework, see Cairney and Weible (2015).

change that goes beyond the existing capitalist mode of production, Buch-Hansen and Carstensen (2021, p. 312) introduce a fourth level of general systemic logic: "A logic is systemic if it is permanent and general: while institutions forming part of a given economic system may change, the underlying logic does not (as long as that system is in operation)". Third-order change does not change the logic of the existing capitalist systems but instead changes the ideas of how a decent and appropriate capitalist system functions—for example, in regard to the role of the government in economic development. A fourth-order change, instead, introduces a different system that abandons key principles of capitalism and represents "much deeper institutional and ideational change" (ibid.). Such an addition is crucial in order to give an outlook on potential existing changes within the eco-social policy discourses of both IOs that might reflect a third-order change but that at the same time silence the more radical change of profoundly decommodifying nature and labour as part of a post-growth policy agenda (Chap. 7).

This book utilises these considerations to interpret the empirical findings. It reconstructs whether the identified eco-social policy discourses can be part of a paradigm or the struggle between paradigms. Referring to the theoretical idea of Polanyi's double movement, and the eco-social policy typology (Chap. 3), the book suggests that there is a dynamic struggle between policy paradigms broadly reflected within green economy and green Keynesianism in both organisations' eco-social policy discourses. This also suggests that there is not necessarily a coherent and unified global eco-social policy paradigm. However, the book proposes to perceive the eco-social policy constellations of ideas promoted by both organisations as a "web of loose [eco-social] ideas" (Von Gliszczynski, 2015, p. 67ff.; see Sect. 7.4).

References

Atzmüller, R., Aulenbacher, B., Brand, U., Décieux, F., Fischer, K., & Sauer, B. (Eds.). (2019). *Capitalism in transformation. Movements and countermovements in the 21st century.* Edward Elgar.

Blyth, M. M. (2001). The transformation of the Swedish model: Economic ideas, distributional conflict, and institutional change. *World Politics, 54*(1), 1–26. https://doi.org/10.1353/wp.2001.0020

Blyth, M. M. (2002). *Great transformations. Economic ideas and institutional change in the twentieth century.* Cambridge University Press.

Buch-Hansen, H., & Carstensen, M. B. (2021). Paradigms and the political economy of ecopolitical projects: Green growth and degrowth compared. *Competition & Change, 25*(3–4), 308–327. https://doi.org/10.1177/1024529420987528

Burawoy, M. (2015). Facing an unequal world. *Current Sociology, 63*(1), 5–34. https://doi.org/10.1177/0011392114564091

Cairney, P., & Weible, C. M. (2015). Comparing and contrasting Peter Hall's paradigms and ideas with the advocacy coalition framework. In J. Hogan & M. Howlett (Eds.), *Policy paradigms in theory and practice: Discourses, ideas and anomalies in public policy dynamics* (pp. 83–99). Palgrave Macmillan UK.

Carson, M., Burns, T. R., & Calvo, D. (2009). Introduction. In M. Carson, T. R. Burns, & D. Calvo (Eds.), *Paradigms in public policy. Theory and practice of paradigm shifts in the EU* (pp. 11–28). Internationaler Verlag der Wissenschaften.

Deutschmann, C. (2019). Karl Polanyi as a theorist of disembedded markets. In R. Atzmüller, B. Aulenbacher, U. Brand, F. Décieux, K. Fischer, & B. Sauer (Eds.), *Capitalism in transformation. Movements and countermovements in the 21st century* (pp. 60–74). Edward Elgar.

Diercks, G., Larsen, H., & Steward, F. (2019). Transformative innovation policy: Addressing variety in an emerging policy paradigm. *Research Policy, 48*(4), 880–894. https://doi.org/10.1016/j.respol.2018.10.028

Fairclough, N. (2010). *Critical discourse analysis* (2nd ed.). Longman.

Foucault, M. (1991). Governmentality. In G. Burchell, C. Gordon, & P. Miller (Eds.), *The Foucault effect. Studies in governmentality* (pp. 87–104). Harvester Wheatsheaf.

Goodwin, G. (2018). Rethinking the double movement: Expanding the frontiers of Polanyian analysis in the Global South. *Development and Change, 49*(5), 1268–1290. https://doi.org/10.1111/dech.12419

Hall, P. A. (1993). Policy paradigms, social learning, and the state: The case of economic policymaking in Britain. *Comparative Politics, 25*(3), 275–296. https://doi.org/10.2307/422246

Jørgensen, M. W., & Philips, L. (2002). *Discourse analysis as theory and method.* Sage Publications.

Lechner, F. J., & Boli, J. (2005). *World culture: Origins and consequences.* Blackwell.

Mendly, D. (2020). Global governance and the double movement. *Global Governance: A Review of Multilateralism and International Organizations, 26*(3), 500–521. https://doi.org/10.1163/19426720-02603006

Merlingen, M. (2003). Governmentality. Towards a Foucauldian framework for the study of IGOs. *Cooperation and Conflict: Journal of the Nordic International Studies Association, 38*(4), 361–384. https://doi.org/10.1177/0010836703384002

Meyer, J. W. (2009). Reflections: Institutional theory and world society. In G. S. Drori & G. Krücken (Eds.), *World society: The writings of John W. Meyer* (pp. 36–63). Oxford University Press.

Meyer, J. W., Boli, J., & Thomas, G. M. (1987). Ontology and rationalization in the western cultural account. In G. M. Thomas, J. W. Meyer, F. O. Ramirez, & J. Boli (Eds.), *Institutional structure. Constituting state, society, and the individual* (pp. 12–37). Sage.

Meyer, J. W., Boli, J., Thomas, G. M., & Ramirez, F. O. (1997). World society and the nation-state. *American Journal of Sociology, 103*(1), 144–181. https://doi.org/10.1086/231174

Miller, P., & Rose, N. (2008). *Governing the present. Administering economic, social and personal life*. Polity Press.

Patomäki, H. (2014). On the dialectics of global governance in the twenty-first century: A Polanyian double movement? *Globalizations, 11*(5), 733–750. https://doi.org/10.1080/14747731.2014.981079

Phillips, N., Lawrence, T. B., & Hardy, C. (2004). Discourse and institutions. *Academy of Management Review, 29*(4), 635–652. https://doi.org/10.2307/20159075

Polanyi, K. (2001 [1944]). *The Great Transformation. The political and economic origins of our time*. Beacon Press.

Princen, S., & Van Esch, F. (2015). Paradigm formation and paradigm change in the EU's stability and growth pact. *European Political Science Review, 8*(3), 355–375. https://doi.org/10.1017/s1755773915000089

Ruggie, J. G. (1982). International regimes, transactions, and change: Embedded liberalism in the postwar economic order. *International Organization, 36*(2), 379–415. https://doi.org/10.1017/s0020818300018993

Sabatier, P. A., & Jenkins-Smith, H. C. (Eds.). (1993). *Policy change and learning: An advocacy coalition approach*. Westview Press.

Skogstad, G., & Schmidt, V. A. (2017). Introduction: Policy paradigms, transnationalism, and domestic politics. In G. Skogstad & V. A. Schmidt (Eds.), *Policy paradigms, transnationalism, and domestic politics* (pp. 3–35). University of Toronto Press.

Streeck, W., & Thelen, K. (Eds.). (2005). *Beyond continuity: Institutional change in advanced political economies*. Oxford University Press.

Surel, Y. (2000). The role of cognitive and normative frames in policy-making. *Journal of European Public Policy, 7*(4), 495–512. https://doi.org/10.1080/13501760050165334

Tag, M. (2013). The cultural construction of global social policy: Theorizing formations and transformations. *Global Social Policy: An Interdisciplinary Journal of Public Policy and Social Development, 13*(1), 24–44. https://doi.org/10.1177/1468018112469805

Turnhout, E., Neves, K., & de Lijster, E. (2014). "Measurementality" in biodiversity governance: Knowledge, transparency, and the intergovernmental science-policy platform on biodiversity and ecosystem services (Ipbes). *Environment and Planning A: Economy and Space, 46*(3), 581–597. https://doi.org/10.1068/a4629

Von Gliszczynski, M. (2015). *Cash transfers and basic social protection*. Palgrave Macmillan.

CHAPTER 3

An Eco-Social Policy Typology: From System Reproduction to Transformation

This book provides an assessment of the OECD and World Bank from an eco-social policy perspective by reconstructing to what degree eco-social policy discourses are promoted by both organisations and what kind of discourses are involved. To have a conceptual baseline for this assessment, this chapter develops an eco-social policy typology based on which policy actors' proposals can be assessed in terms of their eco-social potential.[1] The eco-social policy literature is relatively vague in conceptual terms, missing guidance on how to study policy actors from an eco-social policy perspective (but see also Cotta, 2025). While Matteo Mandelli (2022) has advanced the conceptualisation of eco-social policies, clarifying how they result out of policy integration in explicit attempts to turn trade-offs into synergies, this chapter reviews different policy perspectives that all try to establish these synergetic relationships between environmental and social policy although in part in quite antagonistic ways. The perspectives stem from different literature streams. The review of interconnections and tensions between environmental and social policy draws mainly from policy studies; the review of green growth with green economy and green Keynesianism as two partly opposing perspectives mainly from political economy and environmental economics. Recomposing consumption and production stems mainly from policy studies and political economy, while

[1] An earlier and shorter version of this chapter was published as an article in *Global Social Policy* (Schulze Waltrup, 2025).

R. Schulze Waltrup, *Eco-Social Policy in the OECD and the World Bank*, https://doi.org/10.1007/978-3-031-95429-0_3

post-growth originates in ecological economics. This chapter reviews the different perspectives with respect to their eco-social policy potential and, in doing so, develops an eco-social policy typology on the basis of which an actor's policy proposal can be assessed in terms of its potential to drive a social-ecological transformation. In that sense, it provides heuristic guidance on the question of what kind of policy discourses can be labelled as eco-social and to what degree they entail eco-social ideas and potential. It distinguishes different "degrees of eco-social policy" (Krause, 2021) in terms of their transformative potential and how they enable distinct (political) pathways towards the integration of environmental and social policy. However, the ideal types within the typology are exaggerations and ideal representations of potential reality, which one will hardly find in the empirical realm in the way set out there. Nonetheless, based on the typology, one can investigate the extent to which the empirical material is different from the ideal type, the extent to which there are deviations, and what deviations look like. Chapters 4 and 5 use the typology to reconstruct the eco-social policy discourses that the OECD and World Bank make use of.

To understand the different eco-social policy perspectives and how they refer to both environmental and social policy, this chapter reviews interconnections and tensions between the two policy fields. The reviewing process particularly highlights the divisive tension in the relationships between social policy and economic growth and environmental policy and economic growth, as well as the interconnections between the two policy fields and how they both rely on an intervening government or other public modes of governance that regulate economic developments in terms of their environmental and social impact. Afterwards, the chapter reviews four different eco-social policy perspectives and presents them as an eco-social policy typology, identifying the different ways in which they relate to economic growth and public modes of governance. This typology makes two significant contributions. First, it displays various perspectives and explores the interconnections and relationships between those perspectives. Second, it relates these perspectives to a review of the interconnections and tensions between environmental and social policy, setting the stage for an eco-social policy discussion. Accordingly, this typology examines the different perspectives from various disciplinary backgrounds as part of their role in eco-social policy research.

Reviewing the different perspectives in the order of their transformative potential, the discussion starts with the notion of green growth (not as a perspective on its own but as an umbrella notion covering the two partly opposing perspectives of green economy and green Keynesianism), which assumes that one has to solve the adversarial relationship between social and environmental policy by making economic growth clean and compatible with environmental protection. While the eco-social policy perspective of green economy favours increasing the scope of private actors and market-based solutions in the process of a transition to green growth, green Keynesianism favours increasing the scope of government intervention within the same transition. Green Keynesianism is also more intent on remedying either the ecological consequences of certain social policy constellations or the social consequences of certain environmental policy constellations. The subsequent section presents recomposing consumption and production as a subsequent eco-social policy perspective. The overarching goal of recomposing consumption and production is to favour institutional and structural incentives and cultural contexts that encourage sustainable consumption and production decisions. This entails, for example, channelling resource consumption to social groups in which the utility of resource consumption is highest. This comes along with an appreciation of government intervention and communal modes of possession and consumption. The next section portrays post-growth as the final eco-social policy perspective. Here, degrowth as part of post-growth is an explicit counterdiscourse against the basic assumption of green growth and aims to abandon gross domestic product (GDP) as an indicator of social progress. Post-growth aims to downscale production and consumption patterns to a sustainable level within planetary boundaries while also ensuring comprehensive social welfare provision in that process. This comes along with a strong appreciation of public modes of governance and a strong emphasis on (local) community. The last section of this chapter clarifies how the different eco-social policy perspectives relate to each other and to what degree they either reproduce the given system or entail transformational shifts.

3.1 The Interconnections and Tensions of Environmental and Social Policy

Both social policy, with its welfare state, and environmental policy, with its environmental state, emerged as a result of the ecological and social ruptures brought about by industrialisation and its affiliated political bargaining in the Global North. In numerous countries in the Global North, the welfare state began to take shape in the 1880s with the creation of social insurance programmes, often referred to as the Bismarckian system. It was not until the late 1960s that environmental policy was integrated into state bureaucracies, as governments started to establish environmental ministries and other administrative structures to manage environmental concerns (Duit et al., 2015; Kuhnle & Sander, 2010).

This suggests that the operations of the nation-state's environmental policy are, to a certain degree, dependent on the existing arrangements of the welfare state, including "bureaucratic structures, regulatory modes of interventions, and ideological justification[s]" (Duit et al., 2015, p. 8). As a result, environmental policy has to some extent been layered upon existing established state-structures that have been shaped by social policy arrangements, creating both tensions and interconnections between the various institutional arrangements, regulations and state priorities that underpin both ecological and social policies. In what follows, some of these tensions and interconnections are discussed. This discussion serves to display the overarching context in which eco-social policies are embedded. A particular focus here is on the divisive tension in the relationship between social policy and economic growth, and between environmental policy and economic growth, as well as the interconnections between both policy fields in relying on an intervening government or other public mode of governance that will regulate economic developments in terms of their ecological and social impact.

The evolution of social policy has historically been seen as complementary to economic growth (Meadowcroft, 2005, p. 11), with a generally indifferent attitude towards growth as long as some level of it was achieved. This means that social policy-making has been based on the assumption of a continuous increase in material wealth, a gradually growing population and consistent economic expansion. This steady growth was expected to support stable job and business operations, which in turn would boost tax revenues, thereby funding and expanding welfare services (Gough & Meadowcroft, 2011, p. 495). In other words, economic fluctuations lead to corresponding increases and decreases in tax revenue and social security contributions, which significantly impact the scope of welfare state spending (Büchs & Koch, 2017). This insight is crucial as regards the Global North's ageing populations, which result in increased demand for welfare spending, particularly in the health and pension sector (Bailey, 2015, p. 795). However, social policy is linked to economic growth not only through a one-way connection of dependency; it also creates and drives economic growth (Büchs, 2021a). This is particularly enforced by education policy, family support and active labour market policies (Sakamoto, 2020). Furthermore, unemployment insurance and minimum income schemes support economic growth because they maintain a certain consumer demand in times of unemployment (Büchs & Koch, 2017, p. 30). Moreover, by redistributing income to lower-income households, welfare states ensure a high demand in the economy because it is in these households where the propensity to consume is high. In addition, the welfare state ensures demand through government purchases of goods and services, which are also an essential economic growth factor (Korpi, 1985). In short, economic growth traditionally has a complementary role in the establishment and development of social policy, while social policy has a complementary role to the creation of economic growth as well. This complementary dynamic has traditionally been perceived as a "virtuous circle". However, the sustainable welfare and eco-social policy problem is, as further illustrated below, that this "virtuous circle is actually embedded in a vicious circle of ecological collapse" (Hirvilammi, 2020, p. 6).

Historically, expanding growth has been accompanied by an increase in the use of material resources and an increase in greenhouse gas (GHG) emissions. Ecological economists and earth system scientists highlight that such an increase has led to the transgression of various planetary boundaries (Richardson et al., 2023), because of which environmentalists conclude that in "a finite biosphere, material throughput—resource consumption and waste deposition—cannot grow forever" (Meadowcroft, 2005, p. 11). Accordingly, environmental policy operates under the assumption that there can only be growth that includes such planetary limitations in its conception of development. This is in contrast to the idea of the productivism that characterises traditional welfare regimes and reflects the notion that economic growth is crucial for maintaining welfare state services and traditionally has been rather indifferent to the ecological consequences (Fitzpatrick, 2011, p. 62). Following this assumption, there are two perspectives. One calls for growth to be made congruent with ecological protection (green growth; Sect. 3.2) while the other calls for economic growth to be abandoned as a guiding policy paradigm (degrowth/post-growth; Sect. 3.4).

However, as illustrated above, decreases in economic growth entail a fundamental challenge to existing social policy arrangements and their fiscal sustainability, particularly acknowledging a growing welfare need in light of demographic ageing (Bailey, 2015). Looked at from a global perspective, the tension between ecological and social policy with their different attitudes to economic growth becomes even more pronounced. To decrease existing inequalities between developed and developing countries, at least some growth needs to occur in developing countries so that they become as affluent as developed ones. However, this will lead very likely to increased use of ecological resources. Recent calculations suggest that the global economy would need to grow five times bigger at a minimum to reach the poverty level of Denmark worldwide.[2]

Moreover, social and environmental policies differ in terms of their different cultural perceptions. Hirvilammi and Helne (2014), in one of the first conceptual exploration of eco-social policy, highlight how traditional social policy-making has been culturally blind to certain environmental policy issues. They argue that the traditional anthropocentric thinking enshrined in social policy would blind policy makers to the seriousness of the ecological crisis and make them "prone to favour policy solutions that, due to the severity of the ecological crises are not sufficient" (ibid., p. 2161). This anthropocentric angle reflects various assumptions in policy-making discussions and is deeply embedded within the traditional thinking patterns of (Western) welfare state policy-making that largely relies on the idea of productivism. This way of thinking perceives humans as detached from nature, justifying a one-sided use of nature by humankind: nature possesses a wealth of resources which are available without conditions for the benefit of humans. In that sense, it also perceives humans as superior to other species, which is why humankind can exploit them. Progress is equated with economic growth, and well-being is primarily connected to material prosperity. Furthermore, risks are accepted; people are encouraged to take these risks on the assumption that emerging problems can be solved through technological progress. Finally, this way of thinking pushes economic considerations to be integrated into various decision-making processes and rational logics are considered more critical than intuitive or affective logics. This is exemplified by the traditional importance of economic growth for social policy programmes (Hirvilammi & Helne, 2014).

However, there are tensions between policy fields beyond just economic growth, particularly regarding the (national) scope of

[2] Based on: https://ourworldindata.org/poverty-minimum-growth-needed (accessed 15th of October 2024). This minimum calculation is not necessarily realistic as it expects that economies would simply stop growing when they reached the level of Denmark.

environmental and social policies. While the welfare state is a national construct that provides social policies for its own citizens, the concept of a nationally focused eco-state seems almost contradictory (Meadowcroft, 2005, p. 12). Environmental challenges, such as climate change, ozone layer depletion, air pollution, biodiversity loss, deforestation, desertification, organic pollutants, hazardous waste trading and marine ecosystem conservation, necessitate international cooperation (O'Neill, 2009, p. 34ff.). The dynamic between the Global North and South is also crucial in environmental governance, as the Global North has historically been the largest emitter of GHGs, while the Global South disproportionately bears the consequences (Hubacek et al., 2017).[3] Thus, successful environmental policies require significant international collaboration. Furthermore, while social policy is heavily tied to the services provided by the welfare state, environmental advocates often support self-organising initiatives, where local communities and civic groups manage resources and services, creating closer ties between producers and consumers (Fitzpatrick, 2011, p. 65). This underscores the importance of various forms of public governance that even have the potential to transcend the nation-state as the main organising regulative institution.

Another crucial difference between the welfare state and the environmental state is the means of intervention. The welfare state mainly absorbs income from individuals and businesses and provides public services (for example, education, health and social housing). Taxes and insurance contributions are crucial tools to create revenues while also providing a certain degree of income redistribution. The operations of environmental policies, instead, involve a collection of different measures, such as prohibitions, environmental-specific regulations, environmental standards, taxes and subsidies. Here, environmental policy is redistributive because it redistributes the costs and benefits of ecological services among groups in different regions and at different times (Duit et al., 2015, p. 9; Meadowcroft, 2005, p. 13f.).

How interests are organised also differs. In social policy developments, class interests are crucial. Social policy output has been traditionally

[3] This gap between developed and developing countries is institutionalised in global climate agreements in the principle of "common but differentiated responsibilities and respective capabilities", which was established in the United Nations Framework Convention on Climate Change (UNFCCC) 1992 and remains persistent also in the Paris Climate Agreement (Brunnée & Streck, 2013; Falkner, 2016).

determined by a struggle between a proletarian working class and a class of capital owners. The working class claims the right, for example, to enjoy a minimum of a decent standard of housing, access to health care services, work and long-term financial security. These interests are usually organised through labour parties and trade unions (Gough, 2016, p. 28). This struggle is best captured by power resource theory, which assumes that the extent of distribution policies depends on the distribution of economic and political resources between the different classes (Korpi, 1983). While there is substantial evidence that class struggle significantly influences social policy, it is still important to acknowledge that in certain instances, there are cross-class alliances that implement welfare state arrangements that eventually benefit both workers and capitalist interests. This indicates that, to some degree, there is also a business interest in some types of social policy (Swenson, 2002). In environmental policy, even though there are also profits at stake, this particular class dynamic is somewhat limited. Environmental policy interests are found in more educated and prosperous groups (Franzen, 2003; MacMillan et al., 1997). Moreover, it is predominantly social movement groups that pressure the state to act on ecological problems (Dryzek et al., 2003; Lipschutz & McKendry, 2011). Furthermore, trade unions, which have, undeniably, often played a crucial role in advancing workers' rights, have a somewhat ambiguous stance towards climate mitigation policies, ranging across the spectrum from opposition to support (Thomas & Doerflinger, 2020). Nevertheless, there is a tremendous profit interest at stake in environmental policy. Accordingly, there is usually more vigorous opposition to international climate agreements in countries with large oil reserves. There are also growing business interests which see potential in making profits from the transition to environmental protection technologies. In short, environmental policies will be pressured from both sides of the business, green and brown, but in different ways (Falkner, 2008).

Traditionally, welfare policies were perceived as more relevant than environmental policies since immediate social problems such as disease or poverty were rated more urgent than ecological problems, which are often relatively diffuse and only indirectly perceptible (Gough & Meadowcroft, 2011, p. 494). Furthermore, welfare policies have nurtured interest groups that directly profit from them, which is not necessarily so often the case with environmental policies (ibid.). Moreover, the increasing urgency of investing in mitigation and adaptation policies competes with the urgency of increasing welfare state services. There will be rising costs for climate adaptation policies including resettlement and building climate-resilient infrastructure while there are also increasing demands for welfare services connected with health and pension provision due to demographic ageing. These rising fiscal pressures may lead to fiscal competition between ecological and social policy objectives (Gough & Therborn, 2010, p. 717). Since financial resources are often constrained, there is a danger that social policies will suffer cutbacks due to increasingly progressive climate change adaptation and mitigation policies (Gough & Meadowcroft, 2011, p. 494). Accordingly, climate change has the potential to "displace social policy issues, by providing a new focus of countervailing governance in the twenty-first-century" (Gough & Therborn, 2010, p. 717). This fiscal tension is even exacerbated if one considers the discussion surrounding economic growth and its importance for the general fiscal sustainability of a nation-state. If, for ecological reasons, economic growth is contained without the adoption of additional policies that would create revenue, the fiscal prospects of providing general welfare state services and mitigation and adaptation policies will be undermined.

In addition, economic measures that mitigate climate change and serve ecological protection often have regressive social effects. Carbon taxes and cap and trade schemes have a regressive effect on low-income households, with the result that low-income households will take on a more considerable burden than high-income households if these measures are applied (Büchs et al., 2021). This is due to the fact that the use of home energy stays relatively the same independent of income group, which means that low-income groups spend a larger share of their income on home energy than their high-income counterparts (Dresner & Ekins, 2006; Druckman & Jackson, 2008; Wier et al., 2005). Here, the task of eco-social policy is to ensure that carbon taxes and cap and trade schemes are accompanied by redistributive measures, ensuring that lower-income households do not have to bear the social cost of carbon mitigation policies (Gough, 2017).

Since environmental policy has historically been layered on the structure of the welfare state (Duit et al., 2015), this suggests that there should be notable interconnections and synergies between these two policy fields as part of their operational logic. On a very general level, (progressive) environmental and social policies entail an extension of the state authority to various aspects of the social and ecological life. Both policy fields respond to certain kinds of market failure as well as an insufficiency of voluntary actions on the part of individuals (their steering and organising practices). While social policy serves as a safety net with collective protective schemes and acts as a distributional complement to economic market activities, environmental policy represents a safety net for the ecology and aims to protect against overexploitation. Both policy fields operate under the assumption that market operations and voluntary actions cannot necessarily achieve ecological and social objectives on their own. Therefore, there is a need for a regulating state to assist with this purpose (Meadowcroft, 2005). A notable synergy in that regard is, for example, between climate and health policies. Shifting mobility from car driving to walking and cycling would potentially reduce costs within the healthcare system because it would come with significant reductions not only in GHG emissions but also, due to increased physical mobility, in strokes, cancer, depression and dementia. These synergies are also created by policies that reduce the consumption of animal products (Gough & Meadowcroft, 2011, p. 498). Generally, there are considerable synergies between air quality policies, improving public health and mitigating climate change (Rao et al., 2013).

Environmental and social policies offer some guidance concerning economic development by making specific actions possible or constraining them. As far as social policy is concerned, this entails, for example, modifying incomes and property rights, managing certain economic areas by determining wage levels and enabling negotiations between trade unions and employers. With regard to environmental policies, it covers, for example, planning restrictions, selecting specific technological paths or innovations and altering possible practices of individuals and corporations by taxing or subsidising certain activities and consumption and production patterns (Meadowcroft, 2005, p. 8). However, both policy fields also respond to certain social and political dynamics beyond the economy. From a social policy perspective, this category includes changes in family and class structures and processes of industrialisation and urbanisation; in environmental policy, changes in land use (for example also due to urbanisation), scientific and technological advance and innovation. Since both policy fields respond to similar dynamics, the welfare state can also play a crucial role in managing a just transition, protecting the poorest while transitioning to a low-carbon economy and providing protection against increasing ecological risks as well (Bailey, 2015, p. 795). In environmental and social policy, ecological and social problems do not appear by themselves but emerge in processes of contestation between various actors in the two policy fields (Meadowcroft, 2005, p. 9).

As already pointed out, the establishment of modern environmental policy is a matter of, and partly rests on, ideological justifications of welfare state arrangements (Duit et al., 2015), meaning that both policy fields implement normative considerations in their operations. Depending on one's perspective, these include, for example, equal distribution of essential resources between social groups or of ecological burdens and benefits across time and space. Another example is the principle of justice. In social policy, this can involve the eradication of poverty and everyone having a certain minimum living standard; in environmental policy, the eradication of the burden of ecological degradation on future generations (Meadowcroft, 2005, p. 9).

For the understanding of eco-social policy, this means that their goal is to overcome tensions between these policy fields by operating in a synergetic way while serving social and ecological needs for which the interconnections are the foundation. In other words, eco-social policy characterises environmental and social policy concerns in an integrated way and reconciles tensions between ecological and social welfare systems. The next sections deal with different eco-social policy perspectives that aim to achieve this in different ways. Those perspectives are distinct in their relation to economic growth (core tension between environmental and social policy) and the degree to which they appreciate public modes of governance (synergy in terms of enabling and constraining certain [economic] practices based on government intervention or the organisation of other public actors), both of which have consequences in respect of their potential to drive a social-ecological transformation. This potential and the relationship between the perspectives are further discussed in Sect. 3.5.

3.2 Green Growth

Green growth is the umbrella notion covering the first two eco-social policy perspectives, namely green economy and green Keynesianism. This means that the assumptions of green growth displayed here are adopted by both green economy and green Keynesianism, even though they also entail opposing perspectives. Green growth is appreciative of economic growth and considers it as crucial to solving ecological and social crises. However, green growth proposals often include different and partly opposing frameworks concerning the role of private and public actors. This typology therefore divides green growth into green economy and green Keynesianism (Tienhaara, 2018). Green economy focuses on increasing the scope of private actors while green Keynesianism focuses on increasing the scope of public actors, specifically the role of the government.

Green growth's premise in regard to eco-social policy is that one simply has to overcome the oppositional relationship between social and environmental policy towards economic growth by making it compatible and congruent with ecological protection so that the traditional welfare state dynamic and spending can continue without ecological degradation. While green growth was already being discussed in the 1980s (Colby, 1989), it was only in the 2000s that it acquired traction as an important policy discourse with increasing prominence after the global financial crisis. Generally, green growth refers to a shift from economic growth that correlates with increases in emissions of GHGs and resource throughput to economic growth that has discarded such correlations. It entails four basic assumptions (Gough, 2017). First, growth is generally desirable and is operationalised on the basis of GDP. Second, growth and human well-being are in danger if the ecological foundations of human life are destroyed. Third, there are market failures, including lack of perfect competition and knowledge, detrimental risks and externalities that endanger the ecology. And, fourth, collective corrections are needed to amend these market failures. However, how these corrections should look is hotly disputed, particularly where the role of government is concerned (see green economy and green Keynesianism).

Bowen and Hepburn (2014, p. 409) propose a general definition of green growth. They state that "green growth means increases in economic activity in the long term and possibly short term, without reducing aggregate natural capital". To that way of thinking, increases in GDP usually imply growth of welfare, prosperity and human well-being that are also key social policy objectives (Bowen & Hepburn, 2014). In order to reach that, green growth proposes various changes to the economy, including a shift towards renewable energy resources, recycling waste and other materials, developing new sustainable materials and sustainable harvesting methods, improving (industrial) processes of production and consumption while making the economy into a knowledge-based one instead of one that relies on material throughput in order to grow (Jacobs, 2013). This transition entails two claims. First, green growth emphasises that the costs of changing and preventing economic practices that lead to environmental degradation are not so high as to significantly undermine growth. Second, it assumes that long-term economic costs are much higher if there is no intervention to protect the environment. Generally, green growth claims that environmental protection is a crucial factor for economic growth, alongside traditional factors such as labour, physical capital, technology and human capital (ibid., p. 200ff.). Furthermore, it would be essential to look at nature from a monetary perspective. Accordingly, the environment "provides resources, it assimilates wastes, and it performs various 'environmental services' which sustain life including climatic regulation and ecosystem health" (ibid., p. 204). These services have been traditionally regarded as unpriced. There was no real price incentive to value them—which has resulted in a simple disregard for the consequences of unlimited use of natural capital (Jacobs, 1991). Accordingly, there is a certain underinvestment in natural capital and overinvestment in practices causing ecological degradation and pollution. Green growth argues that if this misallocation is corrected, economic growth could potentially even increase (Jacobs, 2013, p. 204).

Green growth promotes "increasing efficiency" as a crucial theme. It follows the assumption that companies and households fail to achieve energy efficiency due to structural and behavioural constraints. This is because the energy price market also follows the already discussed logic of not taking natural capital into account in its operations by relying on resource extraction and fossil fuels. Accordingly, green growth claims that correcting the market failure of not accounting for natural capital can address structural and behavioural impediments to improving energy efficiency, which can result in reduced emissions, cost savings and increased efficiency within the production process (Jacobs, 2013, p. 206).

Proponents of green growth highlight the "first mover advantage" when advocating green growth policies. The idea is that countries that engage in a sophisticated environmental policy agenda could improve their export capabilities in new environmentally technology-friendly products to other countries (Porter & van der Linde, 1995; Lanoie et al., 2008). Generally, environmental technologies are considered crucial in the transition towards green growth.

Green growth is very much complementary to and has, in many ways, resulted from the discourse of ecological modernisation, which appeared as a policy discourse within Western Europe in the 1980s and has also been discussed in the academic literature as a theory of societal transformation processes (Mol et al., 2009). The ecological modernisation assumption also holds true for green growth (and vice versa). The discourse of ecological modernisation recognises that the process of modernisation has structurally caused the transgression of planetary boundaries.

However, that does not lead to the conclusion that the modernisation project should be abandoned, rather it "assumes that existing political, economic and social institutions can internalize care for the environment" (Hajer, 1995, p. 25). In this process of transforming "the modern project" into an ecologically sustainable one, ecological modernisation especially highlights the role of technology as essential for ecological transformation processes, something that is valued despite the uncertainty of expert knowledge on the cause of and possible solutions to ecological problems (Huber, 2004; Mol & Sonnenfeld, 2000). Ecological modernisation suggests that capitalism does not necessarily have to damage the environment. It stresses that capitalism is dynamic and that ecological issues are key drivers of change. This view also contends that capitalism can facilitate eco-friendly production and consumption, and it argues that alternative economic systems have frequently resulted in worse economic, environmental and social outcomes (Mol & Spaargaren, 2000, p. 23). Ecological modernisation hardly has a revolutionary agenda. It "rather focus(es) on redirecting and transforming 'free market capitalism' in such a way that it less and less obstructs, and increasingly contributes to, the preservation of society's substance base in a fundamental/structural way" (ibid.). This comes with a focus on market processes and economic actors that are assumed to be capable of enacting necessary transformation processes: "Ecological modernisation uses the language of business and conceptualises environmental pollution as a matter of inefficiency while operating within the bounds of cost-effectiveness and bureaucratic efficiency" (Hajer, 1995, p. 31). If earlier ecological protection represented a threat to business, the discourse of ecological modernisation perceives ecological protection as fundamental and a condition for conducting business because ecological degradation would otherwise undermine the condition of doing business. In fact, "ecological deficiency" is taken as a "fundamental impulse that sets and keeps the capitalist engine in motion" (Schumpeter, 1961, p. 83), which can provide a new wave of innovation to reconcile ecological protection and economic growth (Hajer, 2009, p. 83). Thus, being in line with green growth, ecological modernisation entails a conception of changing physical and material flows within the economic system, which reflects "a process of decoupling or delinking of material from economic flows" (Mol & Sonnenfeld, 2000, p. 6).

3.2.1 Green Economy

Green economy follows the general assumptions of green growth as laid out in the previous subsection but has a specific emphasis on increasing the scope of private actors and market-based solutions while rejecting the idea that public actors are crucial in steering green growth developments. While green economy acknowledges that there is a need for some sort of public policy in the green growth transition, it promotes traditional neoclassical economic thinking that public policy should not go beyond guaranteeing basic market operations and modest corrections of market failures. Market corrections are mainly promoted by pricing ecosystem and climate services, internalising them into market cost and valuing them in monetary terms (Tienhaara, 2018). However, while neoclassical economics were practised and promoted largely in the absence of considerations of the ecology, green economy integrates these frameworks into the overarching one. Carbon taxes and emission trading take up a crucial role in that perspective, because if those measures are applied, private actors can still decide on their own how to bring about reductions in GHG emissions (Newell & Paterson, 2010). Carbon taxes raise carbon costs on a general level, effectively increasing the cost of carbon-intensive practices. Emission-trading schemes determine how much an actor is allowed to emit. If they fail to stay within their budget, they must buy credits from other actors who manage to emit less than their limits, allowing them to sell their surplus (ibid., p. 25). This way, cap-and-trade schemes indirectly regulate carbon prices (Gough, 2017, p. 129). The idea behind the distribution of carbon emission allowances with the potential of trading them between actors is that companies would have an incentive to engage in energy efficiency practices as this would result in potential revenue (Newell & Paterson, 2010, p. 95). Comparing carbon taxes and emission-trading schemes, the latter have a stronger ideological fit with market thinking since carbon taxes represent a more decisive intervention by the government into the market. Furthermore, based on emission trading schemes, private and financial actors can create new carbon markets for emission certifications that offer various possibilities of trading them and selling them as various products. Accordingly, carbon markets are perceived as the most flexible and cost-effective solution for private actors to transition to low-carbon practices (Newell & Paterson, 2010).

Green economy empowers private actors and their capacity to act without invoking strong regulations on private actors' practices. It calls, for example, for self-regulating risk management practices by companies. Companies in this view take preliminary and provisional action with regard to ecological protection since they fear possible future public environmental regulations, loss of corporate reputation and legal liabilities should they cause damage owing to neglect of ecological protection practices (ibid., p. 43ff.). Moreover, green economy highlights the entrepreneurial role of finance and the possibility of providing new products connected with ecological protection practices. Accordingly, within finance new segments have emerged, such as providing insurance against weather events, highlighting the importance of adaptation to climate change. While this is not necessarily leading to decarbonisation per se, it supports the process of making people aware of the consequences of climate change (ibid., p. 64). Within finance, green economy also calls on companies to benchmark carbon emissions to provide transparency for investors, who can then decide on their investment strategy. Behind this thinking is the idea that companies that emit more are riskier investments than companies that emit less. This higher risk is due to the potential of regulations being implemented that would aim at reducing resource throughput in production processes, possibly undermining current practices within the corporation. Also, companies with fewer GHG emissions are perceived to have a higher public standing and are less likely to be in litigation (Newell & Paterson, 2010).

Green economy favours market mechanisms to the detriment of government intervention. However, this does not mean that public governance is entirely rejected. For example, in order to have a functioning cap-and-trade scheme, contracts and property rights need to be enforced and accounting systems and methodologies must be devised to measure emissions. Universal rules are also needed on how to measure and calculate emissions to make them comparable. Moreover, a body is needed to monitor emission reporting and decide when trading is possible and who can actually trade. Accordingly, there is a need for specific public governance that acknowledges that carbon markets are somewhat different from other markets as they do not exist as ends in themselves but as means to achieve a societal transition towards a low-carbon economy (ibid., p. 141ff.). Nevertheless, green economy rejects the idea of a strong intervening nation-state that would actively push for "industrial, social and technological advance for the public good" (Goldstein & Tyfield, 2017, p. 2). It argues instead that "[g]overnments are largely restricted to creating the conditions required for green markets to operate (e.g., protection of intellectual property rights, creation of permit trading schemes)" (Tienhaara, 2018, p. 37). This argumentation reflects an appreciation of market-based measures to the detriment of top-down and command-and-control policies. Green economy also promotes vital optimism concerning the development and application of new technologies and innovation, meaning that it encourages and supports society to be reliant on more controversial technologies, such as carbon capture storage (CCS) and nuclear energy production (ibid., p. 43).

In terms of its eco-social potential, green economy is the most modest eco-social policy perspective with the least transformational potential, hardly capturing any profound eco-social policy notions beyond that of greening the economic growth on which traditional social policy-making relies. This would, for example, require an increased awareness of the social consequences of certain environmental policies and the ecological consequences of social policy in green Keynesianism.

3.2.2 Green Keynesianism

Green Keynesianism breaks with green economy's policy prescriptions and argues that solely market mechanisms are inadequate to attain resource conservation and social cohesion in the short and long term. It also assumes that market mechanisms alone are incapable of regulating pollutants that are cumulative and long-term in their effect (Harris, 2013, p. 4). Accordingly, it disregards neoclassical economics largely by calling for comprehensive "interventions by the state comprising both transfer measures and public investments" (Custers, 2010, p. 174). While the green economy perspective, inspired by neoclassical economic thinking, blames government's interventionist role for creating market imperfections, green Keynesianism provides a discursive shift to this perspective, arguing that market imperfections are actually created by the deregulated and free market, or put differently by "'dirty', corrupt, corporate capital and industrial polluters" (Goldstein & Tyfield, 2017, p. 10). Instead, green Keynesianism advocates a comprehensive role for the nation-state in public policy to the detriment of private actors and capital. In this discursive shift, innovators and actors focusing on green and clean technologies are particularly empowered by the government, and social protection is appreciated as crucial within this transition process.

Green Keynesianism also distances itself from neoclassical-economics-inspired monetarism and austerity measures and claims that they have considerable limitations. Accordingly, monetary policies can hardly be targeted at certain ecological investments and fail to target specific economic sectors, demands and jobs. Moreover, traditional monetary policy lacks differentiation mechanisms to distinguish between GDP categories that are ecologically harmful or beneficial. Accordingly, green Keynesians argue that an active fiscal policy is essential to steer the economy towards ecological and social objectives and that this can be a complementary pool to monetary policies of price stability and liquidity (Harris, 2013). Moreover, green Keynesianism claims that austerity measures have detrimental macroeconomic consequences that would create a "paradox of thrift". This paradox states that there is a tendency for private actors to save their savings in times of economic uncertainty, triggering a self-reinforcing circle of not spending, which culminates in economic contradiction (Cooper, 2008). In order to prevent such economic contradiction, the government is supposed to counteract and publicly invest in the economy to ensure a stable demand (Zenghelis, 2012). These state interventions are generally framed as necessary concerning economic inequalities, missing investments in green infrastructure and tackling fossil fuel path-dependency (Harris, 2013, p. 4).

Traditional Keynesianism focused on providing full employment and equal distribution of wealth. While this is also true for green Keynesianism, it adds ecological protection as a critical objective (Cömert, 2019; Keynes, 1936). Accordingly, it focuses on investing in low-carbon sectors and ecologically sustainable activities while also aiming for full employment (Eskelinen, 2015; Tienhaara, 2018). In order to achieve this, green Keynesianism constructs an "entrepreneurial state", which is supposed to invest capital into the innovation and development of green industries and technologies (Goldstein & Tyfield, 2017) as well as cultural, educational and health areas. The overall objective is to abandon the economy's high-carbon path-dependency (Blackwater, 2012, p. 51f.; Harris, 2013, p. 5ff.). Furthermore, green Keynesianism suggests that full employment within a capitalist system is only possible if there is indeed state intervention (King, 2008; Tienhaara, 2018).

Another of green Keynesianism's ideas, following traditional Keynesianism, is to create multiplier effects. Investments lead to an increase in demand as part of a self-reinforcing cycle. Beneficiaries of the investment will spend their benefits on other goods and services which means that groups that have not been directly targeted will receive a benefit from the investment as well. This leads to tax revenues that can be reinvested. The idea of multiplier effects is particularly important to combat economic recession (Blackwater, 2012, p. 53). Such a perspective on public spending rejects austerity measures of the kind that are also often applied in times of economic crisis but, unlike green Keynesianism, promote spending cuts. Here, green Keynesianism argues that public spending cuts would lead to economic shrinkage, decreased tax revenues and a growing deficit, eventually exacerbating crises (Cato, 2013, p. 7). While austerity measures can create economic insecurity, which leads actors to be cautious with their spending, green Keynesianism identifies another problem, namely that the private sector is not investing enough in ecological protection technology due to uncertainties about future environmental regulations. Green Keynesianism claims that the government can control and collectivise these risks and uncertainties by investing in low-carbon industries, which would encourage private actors to invest as well (Zenghelis, 2012, p. 3). This rather short-term crisis response with the aim of promoting the long-term abandonment of high-carbon practices is also essential in Green New Deal proposals. The notion of the Green New Deal is closely linked with green Keynesianism and also mirrors its eco-social potential. Similar to green Keynesianism, the Green New Deal rejects the idea that ecological problems can be solved through the primacy of market mechanisms. Unlike the idea of a free green market, which is closely aligned with the green economy approach, the Green New Deal explicitly aims to improve equity and mitigate inequalities by considering, for example, employment links within its ecological protection policy proposals. In both the Green New Deal and green Keynesianism proposals, market mechanisms take a backseat to other types of policies. When proposed, these mechanisms are closely linked to equity measures, such as revenue recycling for social policy issues. Both approaches emphasise the crucial role of the state in directing a low-carbon transition. In terms of employment, there is a focus on creating jobs that are not only green but also just, with an emphasis on expanding and improving work in the caring sector (Tienhaara & Robinson, 2022).

Nevertheless, green Keynesianism is explicitly not only about crisis management but about conceptualising an interventionist government that, by protecting the poorest, provides long-term stability and socio-economic certainty over the transition to green growth (Bailey, 2015, p. 795; Tienhaara, 2018, p. 149). While the idea of a multiplier effect is usually applied in the short and medium term to mitigate economic crisis, the long-term approach focuses on investments in green infrastructure, technology and social protection. While these investments also create a short-term stimulus by maintaining a certain flow of demand within the economy, their primary objective is to increase its resilience. According to this way of thinking, investments in green infrastructure save the economy from potential ecological pressures and improve its resilience in the long term (Blackwater, 2012, p. 53). Green Keynesianism resonates well with various just transition approaches that highlight the role of the government in steering the transition towards a low-carbon economy and, in that process, supporting disadvantaged groups with social protection measures (Krause et al., 2022). Green Keynesianism calls, for example, for investment in the infrastructure of railways, which are also a labour-intensive industry, providing a path for increased employment. In the field of housing, green Keynesianism means investing in low-carbon housing technologies (Eskelinen, 2015, p. 103). Another long-term challenge is to transition from an economy that relies on waste incineration to a circular economy based on recycling. From the perspective of green Keynesianism, the problem with waste incineration is that it reproduces a reliance on the extraction of new raw materials since used material is not recycled but combusted to produce new energy. However, using recycled material in the production processes is usually more energy-efficient than incinerating waste and extracting new resources (Custers, 2010, p. 183f.). Furthermore, green Keynesianism also calls for the taxation of ecologically damaging practices, such as using landfills or incineration ovens. However, as stated earlier, taxes on their own are not sufficient for green Keynesianism. They need to be invested in public measures promoting recycling or other public policy programmes supporting green infrastructure transitions and social compensation programmes (Custers, 2010, p. 184f.).

In terms of what technologies are supported, green Keynesianism focuses in particular on renewable and energy-efficient technologies, while green economy focuses on technologies such as CCS and nuclear energy. Green Keynesianism adopts a precautionary approach reminiscent of the work of John Maynard Keynes himself, who was strongly concerned with the impacts of uncertainty and how to mitigate them (Berr, 2015). Accordingly, for green Keynesianism, technologies based on CCS and nuclear energy are simply too uncertain and controversial to be relied upon (Tienhaara, 2018).

In terms of its eco-social potential, green Keynesianism, unlike green economy, entails an increased awareness of the need to remedy either the ecological consequences of certain social policy constellations, or the social consequences of certain environmental policy ones. Generally, it follows social policy objectives by protecting vulnerable parts within a "just transition of ecological modernisation" (Kreinin, 2020) that is generally aware of the potentially regressive effects of environmental policy measures and that traditional welfare-state operations, relying on productivism and economic growth, have accelerated the transgression of planetary boundaries. The eco-social potential as well as the relationship of green Keynesianism to different policy perspectives, including post-growth that criticises this inherent productivism of welfare state operations, are further discussed in Sect. 3.5.

3.3 Recomposing Consumption and Production

This section reviews recomposing consumption and production as a third eco-social policy perspective. Its overarching objective is to favour institutional and structural incentives and cultural contexts that encourage sustainable consumption and production patterns. This entails channelling resource consumption to groups who derive the highest utility from it, which is usually the case with low-income households, while preventing excessive consumption by high-income groups (Gough, 2017). It resonates well with green Keynesianism since both policy perspectives presuppose the government as crucial to interventions in market operations. Nevertheless, recomposing consumption and production can also set the stage for post-growth, depending on the discourse's specific empirical configuration. There is a clear tendency to decrease consumption and production levels, thereby reducing resources throughput, as envisaged in post-growth. Furthermore, like post-growth, recomposing consumption and production also promotes collective modes of possession. Accordingly, it hovers somewhere between two stools, relying on green growth but potentially paving the way for post-growth (Gough, 2017; see also the discussion in Sect. 3.5).

3.3.1 Identifying the Problem

Recomposing consumption and production highlights that general green growth ambitions to increase (energy) efficiency in production processes (regardless of the degree of governmental intervention) will not be enough to mitigate tensions between ecological and social protection practices. It argues that existing consumption and production patterns neglect social and ecological protection objectives and must be radically changed to provide comprehensive social policy and ecological protection (Gough, 2017). Two observations are essential to the policy perspective's argumentation. First, recomposing consumption and production observes that inequalities between household expenditure usually increase GHG output, and upper-income groups are usually most responsible for GHG emissions (both at the national and global level) while low-income groups suffer most from ecological degradation (Hubacek et al., 2017; Knight et al., 2017; Millward-Hopkins & Oswald, 2021; Oswald et al., 2020). Second, it observes that countries of the Global North increasingly submit reports of decreasing emissions to the UNFCCC based on territorial accounting, which totals all the emissions produced in a nation-state's territory. However, if emissions are operationalised on the basis of consumption-based accounting (CBA), which includes emissions resulting from the production of imported goods and services while excluding those resulting from the production of goods and services exported to other countries, decreases are only minimal or non-existent (Bows & Barrett, 2010; Nyfors et al., 2020). CBA emissions exceeded territorial emissions in almost all EU countries (except Denmark and Estonia) from 1993 to 2010 (Liobikienė & Dagiliūtė, 2016). Thus, CBA reveals the persistently high proportion of GHG emissions emanating from rich countries. One idea behind CBA is to highlight how the mitigation policies of developed countries might have a potentially positive effect on decreasing emissions in developing countries. Accordingly, recomposing consumption and production policies in the Global North could affect the whole supply chain which would impact production processes in developing countries (Bows & Barrett, 2010). In sum, recomposing consumption and production, by making use of CBA, highlights the persistently high share of GHG emissions from Global North countries as well as the remarkable potential of policies based on its principles to trigger changes in transnational production processes. These two observations, the inequality of emission output and its consequences and the persistent high shares of emissions produced by Global North countries, are core policy problems that recomposing consumption and production refers to. In order to solve them, it aims to shift consumption and production patterns from high-carbon to low-carbon hotspots, which entails improving energy efficiency and steering consumption and production patterns in an equitable way. It also aims at a general reduction of consumption and production levels, potentially paving the way to post-growth (Nyfors et al., 2020).

3.3.2 *Basic Assumption and Corridors*

Recomposing consumption and production acknowledges some scope for individual consumers to change and steer market operations towards green and social product and service provision. Nevertheless, the focus is still on rejecting an individualistic notion of consumer sovereignty as a starting point for policy considerations while assuming that private consumption does not always consider the social and ecological costs of these decisions (Lorek & Fuchs, 2013). From that perspective, public policy needs to ensure that external costs are internalised and visible to individual consumers (Jackson, 2005). The idea of consumer sovereignty also overlooks a significant power imbalance between global retail corporations and individual consumers. The former have much more scope and potential to influence decision-making and the cultural construction of related consumption patterns, a culture that often influences the individual consumer to a quite remarkable extent (Fuchs, 2007; Fuchs & Lorek, 2001). Accordingly, recomposing consumption and production focuses on identifying the cultural and institutional lock-in effects driving ecologically and socially harmful consumption (Lorek & Fuchs, 2013; Sanne, 2002). These lock-in tendencies can differ depending on ethnicity, income, gender and age, which makes a shift to sustainable consumption more difficult for various groups. For example, a rural population can be locked into having a private car for transportation due to an absence of public transportation infrastructure (Seyfang & Paavola, 2008, p. 676).

Generally, recomposing consumption and production favours institutional and structural incentives as well as cultural contexts that encourage sustainable consumption and production decisions for all parts of the population. In order to achieve this, it promotes regulatory policy to shift the consumption patterns of the locked-in consumer, acknowledging that informative and economic incentives are hardly enough on their own to change consumption and production patterns to any large extent. Nevertheless, the upshot is that informative and financial incentives are still a vital part of recomposing consumption and production, but only insofar as they can be supplemented by comprehensive regulation and public procurement policies to foster a sustainable consumption and production culture. It also conceptualises the government as a potential role model for diffusing sustainable practices (Fuchs, 2013; Jackson, 2005; Tukker et al., 2009). From that perspective, government intervention is necessary in order to channel resource consumption to social groups in which its utility is highest. This leads to policies that aim to reduce material consumption by the wealthy since, in these groups, the utility of increased consumption is relatively low compared to low-income groups (Beddoe et al., 2009, p. 2487). Such thinking is particularly well represented in the idea of "consumption corridors".

The general idea of consumption corridors is to create, on the one hand, a minimum and, on the other, a maximum consumption standard. Minimum standards entail that every individual in the present and in future has sufficient resources available to fulfil their needs while at the same time safeguarding social and ecological resources. A maximum consumption standard sets certain limits by preventing and prohibiting excessive, and ecologically and socially harmful, consumption. This maximum standard guarantees that the universal minimum standard is not threatened by excessive consumption by some individuals (Di Giulio & Fuchs, 2014; Fuchs et al., 2021).

Within consumption corridors, there is an important tension. They need to consider and respond to individual notions of having a fulfilled life and universalistic conceptions of human needs which need to be met for everyone. This means that consumption corridors must identify and define (external) conditions that need to be met to live a fulfilled life while also determining at what level consumption not only supports the individual notion of a good life but also damages other people's present and future prospects of fulfilling their needs. In other words, the concept of consumption corridors aims to define a level of sufficiency commensurate with the societal goal of providing every individual with enough for a "good life" without exceeding planetary limitations (Di Giulio & Fuchs, 2014; Fuchs et al., 2021). However, it is essential to note that consumption corridors change over time and, depending on cultural and social contexts, entail different things. Depending on the social and cultural context, there are different satisfiers available to fulfil needs. The satisfiers' ecological and social impact might also change with new technological improvements. Therefore, consumption corridors need to be renegotiated constantly (Di Giulio & Fuchs, 2014; Fuchs et al., 2021). Proponents state that corridors must be determined democratically with civic engagement and participation that includes diverse social strata. Broad participation is crucial for the legitimacy of sustainable consumption corridors and eco-social policy ideas in general (Fuchs et al., 2021; Lindellee et al., 2021).[4]

Consumption corridors require government intervention as they aim to protect the common good, including natural and social resources. This necessitates collective action, often undertaken by national and subnational governments within existing institutional frameworks. Proponents of consumption corridors argue that, just as the government is responsible for protecting against discrimination or harmful behaviour by others, it is also justified in limiting consumption that is significantly detrimental. However, this approach conflicts with concepts of consumer sovereignty and the neoclassical interpretation of individual economic freedom that is emphasised and valued by the green economy (Di Giulio & Fuchs, 2014). Green Keynesianism likewise differs from this approach with its stronger focus on demand control and its close connection to increased levels of consumption and production as a means to ensure full employment (albeit with green conditionalities). The different perspectives are discussed further in a comparison in Sect. 3.5.

3.3.3 Policy Measures

While a consumption or production corridor is a rather broad guiding policy idea, recomposing consumption and production also highlights more specific policy measures that can be well in line with consumption (or production) corridors but can also stand independently. There are two policy options which have already been the object of increased attention from governments. The first entails providing information about ecologically and socially sustainable behaviour, assuming that, if individuals have information about the consequences of certain practices, they will change or adapt them. The second entails economically incentivising sustainable behaviour (Jackson, 2005, p. 121). However, while these policy options form an important part of the recomposing consumption and production perspective, they also underestimate "the rich mixture of cultural practices, social interactions, and human feelings that influence the behaviour of individuals, social groups and institutions" (Stern & Aronson, 1984, p. 2). This highlights the importance of recognising consumption as a social practice and considering the diverse social and institutional contexts

[4] See also on production corridors Bärnthaler and Gough (2023).

in which these practices are embedded (Shove, 2014). Accordingly, next to promoting information provision and economic incentives, recomposing consumption and production particularly highlights regulations and government interventions to shape the cultural and institutional context around consumption and production practices. Most policy measures aim to tackle high-carbon hotspots, such as mobility (in particular aviation and private car use), food systems (in particular meat-rich diets and food waste management) and housing (specifically heating and electronic energy and a general increase in living space in square meters per individual) in specific ways (Nyfors et al., 2020; Tukker et al., 2009).

These regulations also promote synergies with public health, which advocates, for example, banning cars from certain spots in the city, and banning and restricting certain advertising practices, such as advertising meat products, luxury cars and aviation, as well as banning outdoor heating (Nyfors et al., 2020). Furthermore, regulations might aim at greening and socialising supply chains, which might impact the ecological and social protection practices of their suppliers in faraway locations that are not part of the jurisdiction of the initial regulation (Lebel & Lorek, 2008, p. 248f.). Additionally, these policies can entail product standards concerning durability and planned obsolescence, recyclability or efficiency (Jackson, 2005, p. 129). Furthermore, they entail regulations that promote extended producer responsibility by requiring companies to be responsible for their products at the end of their life cycle. This extension of responsibility might lead to the use of resources that are easier to recycle and not ecologically and socially harmful (Lifset, 1993).

Policies that advance certification and labelling call for mandatory transparency provisions with regard to production processes enabling consumers to choose products on the basis of ecological and social criteria. Since they can convey specific information in a compressed manner, labels and certifications are supposed to facilitate sustainable consumer decisions. Moreover, they incentivise corporations to change production processes in order to receive specific labels from third parties (Lebel & Lorek, 2008, p. 253).

Recomposing consumption and production policies undermine the cultural notion of valuing individual possessions, but instead promote modes of communal and shared consumption. This speaks, for example, to establishing product service systems (Lebel & Lorek, 2008, p. 251). In such still quite market-based systems, companies do not necessarily sell products but instead regard the provision of a product as a service. The service provision of products is usually not as ecologically damaging as individual possession of them. One key example here is car sharing. In car-sharing systems, an individual does not rely on having a product as their personal possession but on using the service to phone for a car if needed, which would bring about an overall reduction in car ownership (Tukker et al., 2009). Generally, recomposing consumption and production supports alternative ways of obtaining access to goods rather than individual possession. This broadly speaks to a sharing economy and collectivised ways of owning and using products and services. These approaches might limit the impact of "positional goods" that are usually private individual possessions and have the effect of reducing positional-goods-driven consumption competition (Hirsch, 2005). This means that recomposing consumption and production calls for a broad expansion of the public sector to provide public goods and services, following the thinking that some goods should be held in common because they are valuable for everyone or have been created by society as a whole (Beddoe et al., 2009). Such public policies would entail, for example, the provision of public beaches, green areas, public baths, community washing centres or collective and decentralised electricity generation (Nyfors et al., 2020; Sanne, 2002; Tukker et al., 2009). In terms of regulating marketing practices, recomposing consumption and production advocates the placement of warning labels on products, similar to those used for cigarettes. It also calls for comprehensive taxes on advertising practices and supports the notion of ethical marketing in creating a sustainable consumption culture. Ethical marketing states that marketing must not provide misleading information or exploit brand similarities since this would undermine transparency and produce consumer confusion. It also rules out marketing directed at children because it would condition children to the idea that material consumption is a way of showing appreciation in (immaterial) family relationships (Schor, 2004). In short, ethical marketing calls for marketing to be focused only on people who can understand the persuasive and commercial logic of marketing strategies and objects to an appreciation of personal relationships based on material consumption (Lebel & Lorek, 2008).

As indicated above, alongside the reconfiguration of consumer culture and related government regulations, there is still a persistent use of economic instruments to steer consumption and production patterns. These instruments include fees or taxes on high-carbon practices, nudges, subsidies for low-carbon practices and personal carbon trading (PCT) schemes.

Economic instruments are particularly effective when alternative practices are available. These alternatives are supported to the disadvantage of harmful behaviours, leading to an increase in beneficial behaviour (Nyfors et al., 2020; Tukker et al., 2009). De Camillis and Goralczyk (2012) propose a smart value-added tax (VAT) with a differentiated rate depending on the product's ecological and social impact. A smart VAT rate would be connected to the life cycle assessment method of differentiating VAT rates. Life cycle assessments assess the ecological impact of a product using quantitative methods. Such assessments investigate the product's complete lifecycle from extraction to disposal or reuse, monitoring various factors such as use of energy and resources, harmful releases to soil and water and GHG emissions. Coupling VAT rates to life cycle assessments would standardise the process of differentiating VAT rates depending on the product's ecological and social performance. Generally, smart VAT of this kind represents an incentive to consume ecologically and socially beneficial products to the detriment of ecologically and socially harmful ones (De Camillis & Goralczyk, 2012). However, smart VAT would also need to take account of the potential regressive effects of VATs (Thomas, 2020). PCT schemes distribute GHG emission rights at the household and individual level. Such schemes would be mandatory, and citizens would receive a certain number of allowances that would be calculated based on emission reduction targets. Allowances are tradeable and create a new market in which citizens with lower emission output can sell their allowances to citizens with a higher emission output that exceeds their limit. Depending on reduction targets, the overall total of budgets is often reduced within a certain period (Fawcett & Parag, 2010). PCT schemes complement other measures to reduce GHG emissions. They operate like a financial incentive for individuals to make the best of existing programmes or labels. The general goal is to make citizens more aware of their energy use and financially incentivise emission output reductions through the potential to sell surplus allowances to other individuals. PCT is quite similar to existing cap and trade schemes, such as the EU Emission Trading System, but it is also unique in regard to the substantial public involvement because it entails every individual or household participating in the scheme. Proponents argue that this empowers the public (ibid., p. 334). It has been found that PCT schemes are generally progressive, imposing a greater burden on high-income than on low-income groups. Low-income groups can benefit from their surplus allowances by selling them. However, it remains crucial to protect individuals within low-income groups, particularly those with high mobility needs or more children, by allocating more allowances to them to ensure distributional fairness (Pitkänen et al., 2022).

Another element of recomposing consumption and production is "nudging". This is about providing citizens with a specific "choice architecture" that would make ecologically and socially beneficial consumption practices more favourable and easier (John et al., 2009). "Nudges" are interventions that acknowledge that citizens are characterised by bounded-rationality in view of the constraints on processing all available information and being aware of all the consequences of one's practices. Accordingly, individuals do not necessarily see all possible options within decision-making processes and often cannot know what the apparent best decision is. They may also not be eager enough to find the best decisions. Decision-making is much more "driven by habits of thought, rules of thumb and emotions" (John et al., 2009, p. 363). Nudges acknowledge this and are designed to lead individuals automatically to a favourable and generally desired decision. However, to trigger large-scale systemic change, nudges are rather inappropriate since they consist of mechanisms to bring about shifts in small individual choices within the current system (John et al., 2009). Examples include the provision of a vegan or vegetarian option in public canteens as the default option or offering fruit instead of sweets at children's eye level in supermarkets (Nyfors et al., 2020; Thaler & Sunstein, 2008), which have to do with improving public health and, in the case of vegan food, reducing the ecological impact of food choices.

Furthermore, recomposing consumption and production highlights the government's implicit and symbolic power to steer consumption and production processes and to shape consumer culture. Governments display "in sometimes subtle but very powerful ways the kinds of behaviours that are rewarded in society, the kinds of attitudes that are valued, the goals and aspirations that are regarded as appropriate, what success means and the worldview under which consumers are expected to act" (Jackson, 2005, p. 130). Accordingly, government-led policy signals can have a remarkable impact on norms and cultural expectations. Nevertheless, for a government to successfully create a sustainable consumption and production culture, there should be no inconsistencies or hypocrisy in its practices. If the government does not operate in accordance with the tenets of its sustainable consumption strategy and policy, it undermines the credibility of government-led information campaigns. If the government manages to show that it is serious about enacting and demanding sustainable practices, individuals and businesses are more likely to change their practices. Also, the government can identify potential problems that arise in the transition to sustainable practices and will, therefore, be able to work on solutions to these problems proactively. Moreover, if the government as an employer encourages and practices sustainable behaviour, this may also affect their employees' practices in their homes (Jackson, 2005). By including ecological and social criteria in the procurement of goods and services, a government can show consistency between talk and action in regard to its sustainability policy. Furthermore, sustainable procurement policies send a demand further up the supply chain and support the creation of a market for sustainable goods and services. According to this way of thinking, as indicated above, the government represents a symbolic role model for the desired behaviour of its citizens and might trigger new market creations. In short, public procurement policies send important symbolic signals to the broader population about what kind of consumption is appropriate and desired (Tukker et al., 2009). Examples of sustainable public procurement include providing organic and fair-trade food while decreasing meat consumption in public canteens, providing electric cars in public car fleets and only procuring certain certified and labelled products and services (Tukker et al., 2009; Wahlen et al., 2012).

3.4 Post-growth

This section presents post-growth as a fourth eco-social policy perspective. Post-growth is here understood, following Gerber and Raina (2018, p. 353), as a "combined application and theorization of degrowth, agrowth, steady-state economics and post-development". Broadly speaking, this represents a policy agenda of transitioning towards a society with a smaller resource metabolism, which is captured under the term "degrowth", in particular for the Global North. This usually aims to reach a steady-state economy in which production and consumption patterns are downscaled to be within planetary boundaries. The perspective of agrowth acknowledges that if such a steady-state economy operating within planetary boundaries were to be reached, there could be economic growth as long as it did not lead to transgressing ecological and social objectives. It also acknowledges that for some countries at a certain time or in a particular region, it can make sense to pursue growth (the Global South), but for others, it does not (the Global North). This acknowledges that developing countries still need to grow to a certain extent. It also takes account of post-development studies that promote alternative conceptions of development that are not based on the Western hegemony and its "imperial mode of living" (see Sect. 7.2). This present section often refers to degrowth as part of post-growth. Nevertheless, notions of agrowth and post-development studies also play a crucial role (in particular with regard to the Global South). As shown in the following subsection, degrowth argues that abandoning GDP growth is inevitable to enable humanity to reach a safe space within planetary boundaries. As a solution, it proposes to downscale production and consumption patterns to a sustainable level of not exceeding planetary boundaries and eventually to reach a steady-state economy. In that process, comprehensive and far-reaching welfare policy measures will be implemented to maintain reasonable levels of well-being. From a post-growth perspective, those measures can be part of various policies, such as community currencies, non-monetary exchange systems, alternative credit institutions, redistributive policies, such as income and capital caps, work time reduction, a citizen's income, new ownership patterns, investments in public security and public goods and redistributive taxation system (Sect. 3.4.3). Post-growth resonates well with elements of recomposing consumption and production that focus on decreasing consumption levels by reducing excessive consumption on the part of higher-income households. Nevertheless, while recomposing consumption and production can also go well with increasing economic growth, degrowth explicitly abandons the policy trajectory of economic growth, at least for the Global North (Sect. 3.4.2).

3.4.1 Problematising Green Growth and GDP

Degrowth rejects green growth's assumption that *necessary* reductions in energy and resource throughput to reach the safe space within planetary boundaries can increase economic growth. Accordingly, it is an explicit counterperspective to the notion that decoupling economic growth and resource throughput is feasible. Degrowth emphasises that there is no historical evidence for absolute decoupling of economic growth and resource throughput and observes that models indeed project that it cannot be achieved even under favourable conditions (Hickel & Kallis, 2020; Vadén et al., 2020, 2020). Concerning resource throughput, global trends indicate some evidence for relative decoupling but not for absolute decoupling. Instead, there is some evidence that recoupling takes place. The share of services in global GDP has increased from 63 per cent in 1997 to 69 per cent in 2015 while, at the same time, the material footprint has accelerated, even outpacing GDP growth. This observation also holds true if we look solely at high-income countries with state-of-the-art technology (Hickel & Kallis, 2020). There may be various reasons for this. First, and unexpectedly, the provision of services might necessitate a higher resource input than the production processes of goods. Second, the revenues of the services might be used to consume resource-intensive goods (Kallis, 2017). Third, the first (raw materials) and second (manufacturing) sectors' need for resource consumption might have increased to such an extent that it outpaces all possible effects of increasing service provision (third sector). However, whatever is the case, this points to the argument that there is no empirical evidence that simply switching from goods to services will automatically reduce resource throughput in the global economy (Hickel & Kallis, 2020). While a circular economy, an idea also often put forward in green growth proposals, could, in principle, reduce resource throughput to some extent, there is still a significant share of the latter that cannot be recycled because 44 per cent are food and energy inputs, and 27 per cent are due to building and infrastructure stocks (Haas et al., 2015). Moreover, as Ward et al. (2016) point out, efficiency cannot increase forever since there are simply physical limits and when these limits are reached, increasing GDP drives resource use up again. Most studies estimate the decoupling of economic growth and resource throughput based on an assumption of 2 to 3 per cent growth in global GDP. However, decoupling becomes more likely and is more profound if the growth rate is assumed to be zero or less than one per cent, but this would run against projected pathways and existent green growth narratives (Hickel & Kallis, 2020).

In the field of emissions, as distinct from resource throughput, absolute decoupling is achieved by transitioning to renewable energy production. Empirically, there is a steady trend towards relative decoupling. However, decreasing emissions to achieve climate targets is not solely about reducing the flow of emissions but instead keeping emissions under a budget, which is a matter of stocks. Then the question is not primarily whether we can achieve absolute decoupling but how fast we can decouple to stay within the targets of the Paris Climate Agreement of 1.5 and 2 degrees (ibid.). Even models with optimistic assumptions (significant improvements in efficiency and sophisticated climate mitigation policies) state that it is unlikely that decoupling will take place fast enough to stay within carbon budgets of 1.5 or 2 degrees if the global economy continues to rely on economic growth. The pressure for continued economic growth outpaces carbon budgets since, with increased economic growth, there is usually more energy demand (Raftery et al., 2017; Schröder & Storm, 2020). Models that indicate that green growth with the objective of 1.5 or 2 degrees would be possible rely on CCS technology. However, the functionality of CCS still needs to be proven, and there are many insecurities attached to these technologies (Hickel & Kallis, 2020).

Acknowledging this, the precautionary principle is deeply embedded within post-growth. It takes the perspective that we should assume certain economic limits instead of hoping that unrealistic expectations of technological improvement might still be realised in the future (Kallis, 2011, p. 874). Nevertheless, in calculations similar to those that estimate the relationship between economic growth and resource throughput, the feasibility of reaching climate targets changes if the GDP baseline is changed. If the global GDP growth rate is lowered to 0 to 0.45 per cent and coupled with high efficiency and comprehensive mitigation policies, the 2 degrees objective will become feasible (Schröder & Storm, 2020). Overall, existing evidence suggests that simply increasing efficiency is not enough and that reductions in aggregate economic activity must accompany these reductions. This is what post-growth is trying to accomplish while maintaining a reasonable level of well-being in the process.

In explaining the shortcomings of green growth, post-growth highlights the rebound effect (Kallis, 2011; Schneider et al., 2010). The rebound effect states that if the costs and the ecological impact of products are reduced, it can create rebound effects to the detriment of overall ecological performance. First, there is an income effect, meaning that an increase in income due to cheaper and more efficient products leads to more income becoming available for other potentially energy-intensive goods and services. Second, secondary effects entail that there will be increased demand for improved and often cheaper products due to increased efficiency. Third, quantity adjustment means that energy efficiency reduces the need for fuel or energy services. Consequently, these prices will go down, potentially leading to more purchases of these elements (Hertwich, 2005). Moreover, due to lower energy prices, companies will also potentially produce more supply (Greening et al., 2000, p. 391). The rebound effect is particularly strong with respect to luxury products, which are often energy-intensive and socially exclusive (such as trips in aircraft), as consumers tend to spend more on luxury goods as soon as they become cheaper (Hertwich, 2005, p. 93).

In addition to problematising the ecologically detrimental impact of increasing GDP, post-growth also problematises GDP's shortcomings in measuring human well-being, a traditional social policy focus. One issue is that GDP may also increase if there is an increase in expenditure due to practices that involve losses in welfare, such as crimes or road accidents (Costanza et al., 2015, p. 133). Moreover, GDP fails to incorporate domestic (care) work, often carried out by women (Waring, 1988), which is fundamental for social reproduction but often unpaid. It also fails to incorporate the value of enjoying common goods (Kallis et al., 2018, p. 294). An economy that is increasingly shaped by marketisation is also characterised by rising GDP, but marketisation is also fairly indifferent to welfare outputs (Stiglitz et al., 2010, p. 49). In addition, GDP does not say anything about how prosperity is distributed within society, which suggests that high GDP can go along with high inequality (ibid., p. 44). However, if inequality has a negative effect on well-being (Wilkinson & Pickett, 2009), increases in GDP might be accompanied by losses in well-being if they bring with them an increase in inequality (Büchs & Koch, 2017, p. 46f.; Douthwaite, 1999, p. 17). Also, other indicators that incorporate ecological and societal costs have been stagnating despite GDP growth (Beça & Santos, 2010; Kubiszewski et al., 2013). While, traditionally, post-growth proponents referred to the Easterlin Paradox, the observation that once a certain level of GDP has been reached in a country, an increase in it is no longer accompanied by an increase in the subjective well-being of the population (Easterlin, 1974; Schneider et al., 2010); the Easterlin Paradox received some criticism recently (for an overview see Büchs & Koch, 2019, p. 156f.). Important aspects of this criticism are, for example, methodological issues such as that GDP in principle can rise indefinitely but that well-being is measured by bounded scales (Büchs & Koch, 2017, p. 61). Also, studies that focus on the relationship between well-being and the transgression of planetary boundaries (which correlates with GDP) observe that countries with higher GDP perform better in terms of need satisfaction, which undermines the validity of the Easterlin Paradox to a certain degree (O'Neill, 2015; Fanning et al., 2022).

The post-growth perspective also highlights that ecological protection and sharing practices go along with increases in well-being (Kasser, 2017), and that the more equally incomes are distributed, the happier people are (Wilkinson & Pickett, 2009; Reyes-Garcia et al., 2016). Further, it highlights that existing provisioning systems—a "set of related elements that work together in the transformation of resources to satisfy a foreseen human need" (Fanning et al., 2020, p. 3)—fail to efficiently moderate the relationship between physical and social systems (O'Neill et al., 2018). The more fundamental human needs a country succeeds in satisfying, the more planetary boundaries are usually transgressed. Wealthy nations fulfil most human needs but transgress planetary boundaries. Some developing countries hardly transgress any planetary boundaries but fail to fulfil various human needs (O'Neill et al., 2018). Here, degrowth claims to represent a provisioning system that manages both to satisfy basic human needs and to stay within planetary boundaries or what Raworth (2017, p. 44) has called the "safe and just space for humanity". In short, degrowth assumes that well-being would increase if reduction in economic throughput (potentially an inevitable loss of GDP) is accompanied by a cultural shift emphasising redistribution, ecological protection and sharing (Kallis, 2018). The following subsection discusses post-growth's and degrowth's basic assumptions more extensively.

3.4.2 *Basic Assumptions of Post-Growth*

Before discussing several basic assumptions of post-growth with a particular focus on the degrowth literature, this subsection first asks what degrowth actually means and what its characteristics are. Schneider et al. (2010) define degrowth as "an equitable downscaling of production and consumption that increases human well-being and enhances ecological conditions at the local and global level, in the short and long term" (Schneider et al., 2010, p. 511) Another quite similar definition is provided by Jason Hickel (2020, p. 2), who describes degrowth as a "planned reduction of energy and resource throughput designed to bring the economy back into balance with the living world in a way that reduces inequality and improves human well-being". In light of the assumed bad prospects for green growth's success in staying within planetary boundaries (previous subsection), a more comprehensive reduction is said to be an inevitable necessity that may come along with losses in GDP (Buch-Hansen, 2018; Kallis, 2011). It is important to note that degrowth is not about reducing GDP per se but instead about decreasing resource and carbon throughput, which is in line with ecological protection practices, while at the same time ensuring the provision of well-being. Such a decrease is likely to reduce GDP (in the Global North). Degrowth emphasises that one has to prepare for that on an institutional level so that these reductions do not come with declines in well-being. Such preparations and transitions are what degrowth aims to accomplish (Hickel, 2020).

The focus on economic growth in policy-making reflects various problematic political processes that damage human and ecological well-being. These processes entail, for example, "the commodification of commons, and the appropriation of human labour and natural resources" (ibid., p. 3). Moreover, degrowth assumes that the economic growth regime, as it currently operates, negatively affects well-being due to increased status competition and increased pressure to perform productively, causing, for example, mental health problems. It highlights that *individual* growth would weaken *communal* relationships due to individualised market relations. It also states that economic growth fails to create trickle-down effects to eliminate poverty (for a review see Büchs & Koch, 2019). In short, degrowth aims to establish a different set of cultures and institutions, which then mitigate and prevent potential losses of well-being if economic growth declines (Büchs & Koch, 2019, p. 3). However, it does not claim that degrowth should be the permanent direction of policy-making, instead highlighting current detrimental patterns of economic growth that cause ecological degradation while at the same time failing to distribute growth equally. It simply assumes that human progress and increases in well-being are possible without economic growth (Schneider et al., 2010). Similarly, degrowth does not mean unconditional shrinkage. Certain sectors, such as renewable energy, shared transportation systems and poor regions, still need to grow selectively (Schneider et al., 2010, p. 512). Hence, broadly speaking, degrowth aims to reduce the human impact on the ecology, redistribute income at the national and global level and shift from a materialistic cultural notion of well-being to a participatory and convivial one (Cosme et al., 2017, p. 2).

One goal of degrowth, in line with the definition's focus on limiting resource and economic throughput, is to establish a steady-state economy, which has a "constant stock of capital, maintained by a low rate of throughput that is within the regenerative and assimilative capacity of the ecosystem" (Daly, 2008, p. 3). Such an approach calls for developed countries to establish a steady-state economy while acknowledging that developing countries still need to grow to achieve well-being (Kerschner, 2010). Daly (1992) states three general policy objectives for a steady-state economy: (1) sustainable scale of resource use, (2) fair distribution of income and (3) efficient allocation of resources. All presuppose sophisticated and comprehensive measures of regulation and redistribution (Cosme et al., 2017). Daly and Farley (2011, p. 55) suggest that the rate of extraction of non-renewable resources should not exceed renewable energy creation. Nor should emissions exceed nature's capacity to absorb them. A stable or shrinking population is desirable from a degrowth perspective as this would lower pressure on planetary boundaries, whereas from a green growth perspective, population growth is desired since it is a driver of economic growth (Gough, 2017, p. 177). Crucial for degrowth is to abandon the inherent productivism of industrialised economies and instead focus on different non-materialistic ideas of well-being that promote a life independent of large incomes and luxurious consumption. Instead, the focus is on nurturing relationships with other people and their community and self-development (for an overview, see Büchs & Koch, 2017, chapter 5). This entails, for example, ideas of eudemonic well-being (Ryan et al., 2008), human flourishing and capabilities (Nussbaum, 2003; Sen, 1999), "joy of living" (Schneider et al., 2010, p. 513), the "good life" (Muraca, 2012), "voluntary simplicity" (Elgin, 1982) or "alternative hedonism" (Soper, 2008), which are all connected to an alternative image of prosperity independent of economic growth and consumption (Cassiers, 2015; Jackson, 2009). To achieve this, the notion of "care" is put at the centre of degrowth, emphasising that valuing care work will lead to a more fair and just society as far as the distribution of work and well-being is concerned (D'Alisa et al., 2015). In addition, it focuses on the "non-chrematistic value of local, reciprocal services" (Schneider et al., 2010, p. 512), which is, as discussed in the previous subsection, often not captured by GDP. This is also described as the "care economy" of degrowth. All social provisions and goods which are crucial for sustaining human life are therefore put at the centre of the economy, covering care work, time, wisdom, experience, knowledge, skills, love, empathy, reciprocity, teaching and learning, which can all be exchanged but do not have a specific monetary value (Barry, 2012; Coote & Goodwin, 2010, as quoted in Gough, 2017). This entails a significant reallocation of care work (both paid and unpaid) responsibilities between genders, along with a shift away from treating care activities as market commodities (Dengler & Lang, 2022).

As far as the Global South is concerned, post-growth argues that developing countries do indeed need to grow, with increases in resource and energy throughput to fulfil basic needs. Accordingly, degrowth focuses on changing policies in countries that exceed their planetary footprint per capita—which are mainly countries of the Global North (Kerschner, 2010). Furthermore, degrowth highlights the fact that the consumption of high-income countries is to a large degree (up to the equivalent of 50 per cent of their consumption) based on a net appropriation of resources from other parts of the world, indicating that the ecological impact of high-income countries' consumption is experienced mainly in other parts of the world (the Global South). Acknowledging this, degrowth reflects the notion of decolonisation and preventing the Global North from appropriating the fair share of the atmospheric commons and natural resources that should belong to the Global South (Hickel, 2020, p. 5). However, post-growth, in general, is quite critical of the narrative of "GDP growth against poverty" for the Global South, as this approach has usually come with quite unsustainable growth in ecological, social and financial terms in the Global South as well (Gerber & Raina, 2018). Instead, a post-growth perspective on the Global South focuses on direct need satisfaction based on specific local conceptions of "the good life" and wealth distribution. Accordingly, the Western-inspired lifestyle of the ruling classes in developing countries would also need to be changed, given that it is strongly tied to the ruling class of the Global North (Demaria et al., 2013; Gerber & Raina, 2018). In India, for example, high growth rates came along with high rates of inequality that drive the lifestyle of the upper classes while the majority of the Indian population lives in poverty (Agrawal, 2016). Acknowledging this, post-growth aims at increasing the well-being of everyone instead of accelerating an unsustainable lifestyle for the few in the Global South. Objectives should be the democratically defined targets of a social-ecological lifestyle, which would, in the end, also be about redistribution, since there is actually a considerable amount of wealth in the Global South, though it is only held by a few (Gerber & Raina, 2018).

Post-growth policies are particularly characterised by public modes of governance, emphasising the role of collective decision-making and community (Gough, 2017; Koch, 2020). Private actors are still crucial in providing innovation regarding energy efficiency, but because there will be many difficult-to-quantify benefits, such as climate adaptation or maintenance of ecosystems, the public sector will play a crucial role. In order to achieve a cultural and institutional transition, post-growth proposes various policies such as community currencies, non-monetary exchange systems, alternative credit institutions, redistributive policies such as income and capital caps, promotion of work time reduction, a universal basic income, collective ownership patterns, investments in public social security and public goods and redistributive taxation systems (Cosme et al., 2017). The next subsection discusses various policy proposals in depth.

3.4.3 Post-Growth's Policy Proposals

This subsection discusses some policy proposals typical of post-growth, focusing first on policy proposals concerning the labour market, such as working time reductions and job guarantees before discussing local currencies and alternative monetary exchange services in addition to alternative ownership patterns instigating economic democracy. Finally, it discusses redistributive policy ideas, such as a universal basic income and caps on income and wealth.

Post-growth proposes to mitigate possible pressure on the labour market and employment due to decreases in economic growth and a transition towards a low-carbon economy through a reduction of working hours and a specific configuration of job guarantees (see, for example, Alcott, 2013; Cosme et al., 2017; Gough, 2017; Kallis et al., 2013; Schor, 2015; Unti, 2015). Reduction of working hours has the potential (especially in the short term) to reduce unemployment and is thus crucial to mitigating potential employment losses due to a decline in economic growth. Moreover, it can improve individual well-being, although the ecological protection impact is dependent on other policies ensuring that any free time that becomes available will not be devoted towards ecologically damaging practices (Kallis et al., 2013). The reduction of working hours has a tremendous social policy impact. It enables more people to work if they want to work, people to work less and have more leisure time, to take up more meaningful activities or to care for nurturing relationships. For advances in personal well-being, health, education, meaningful social relationships and the quality of the surrounding ecology are particularly important. A reduction in working hours would create the necessary spare time to work on these. Moreover, there might be less work-related stress, improving mental well-being (Kallis et al., 2013, p. 1558). Reducing working hours can also encourage the integration into the labour market of individuals who are under certain time restrictions as a result of non-enumerated care practices, for example, women or care providers (Kallis et al., 2013, p. 1557). Work time reduction proponents also emphasise that it is usually households that are short of time who use more time-saving devices which usually have a higher ecological impact (Devetter &

Rousseau, 2011). But more leisure does not automatically lead to less resource use. It needs additional recomposing consumption and production policies to ensure that preferences are changed to bring about a greater appreciation of "soft" activities, such as reading, playing, spending time with friends or investing in education. Hence, the relative prices of sustainable practices must be less than those of unsustainable goods and practices (see also Sect. 3.3). Post-growth calls for a reduction of working hours while the salary stays the same, since this would also be a crucial redistributive mechanism from capital to labour. The thinking here is that labour would directly benefit from technological gains in productivity by having more time for leisure (Kallis et al., 2013, p. 1561). However, to be successful, reductions in work time would need to be accompanied by sound investments in the social security system and profound shifts from taxation of labour to taxation of capital (Spangenberg, 2010).

Post-growth aims to supplement work time reduction policies with a job guarantee (Alcott, 2013). This follows from the idea that employment should be considered not only from an economic perspective but rather as a political and social right. Given social and ecological objectives, the jobs in question would mainly cover ecological protection and communal practices, but they could also be part of production processes. They would be particular work arrangements that were not supposed to be dependent on economic growth. Crucial for a post-growth perspective is to acknowledge that there is a certain social value attached to work and to appreciate that work is crucial to participation in society. Nevertheless, a job guarantee cannot replace the general social safety net since there will also be people who cannot be forced to work, people who are unable to work, or people who, due to their personalities (for whatever reasons), are unsuitable for work (ibid.).

Post-growth envisages an entirely different way of organising money and criticises how the existing monetary system depends on and reproduces economic growth. Most money in circulation is created by private commercial banks via debt and disappears out of the money flow if the money is repaid, meaning that most money is in flow because someone (from an individual to an organisation or a nation-state) incurred a debt. This method of money creation is called fractional reserve banking (Johanisova & Wolf, 2012). But because most banks are restrained by their shareholders' interest in increasing profits, they will not necessarily invest in social and ecological protection enterprises but ones that provide the greatest profit margin but do not necessarily create social benefits for the common good. This approach to money creation also brings, from a post-growth perspective, the problem that money circulation depends on increasing debt and increasing economic growth (to repay debts). If actors stopped taking on debt while at the same time paying off their debts, the circulation of money would decrease (Johanisova & Wolf, 2012, p. 565). There are various proposals that go beyond this organisation of fractional reserve banking, such as phasing out money creation by private banks and instead demanding that governments use their central banks to bring debt-free money into circulation (Huber & Robertson, 2000). Other proposals incorporate creating local currencies and accordingly a plurality of valid currencies. The latter scheme is particularly encouraged by post-growth (see, for example, Buch-Hansen, 2014; Cosme et al., 2017; Dittmer, 2013, 2015; Kallis, 2011; Latouche, 2009; Schneider et al., 2010). Dittmer (2013) suggests four basic motivations for establishing local currencies: First, local currencies improve community building and social networks. This entails encouraging "the production of relational goods, such as friendship and neighbourliness" (Latouche, 2009, p. 70), which follows the objective of promoting conceptions of well-being that are operationalised on the basis of local community relationships and not individual material possession. Second, local currencies aim to promote alternative values of economic exchange, potentially changing consumption culture. Third, they have the potential to facilitate alternative livelihoods potentially abandoning the necessity for formal employment and emphasising self-determination in production processes. Fourth, they aspire to "eco-localise", meaning the localisation of production and consumption processes to protect the ecology and improve local social networks. Usually, four different types of local currencies are discussed: local exchange trading systems or schemes (LETS), time banks, time-based (HOUR) currencies and convertible local currencies (for more on this, see Dittmer, 2013).

As far as an equal distribution of economic power is concerned, post-growth is critical of the fact that the power of multinational corporations, acquired thanks to ongoing and decades-long deregulation policies, undermines economic democracy as well as the general population's potential to influence economic practices (Johanisova & Wolf, 2012, p. 564). In order to establish more egalitarian economic decision-making, post-growth encourages workers to hold shares in their company, which makes them part of the governance structure and incorporates them into decision-making. However, this must not lead to a "one share–one vote" system, which would undermine egalitarian decision-making because individuals with more shares would have more power in the process. Moreover, the possibility that shares can be traded depending on their value, which usually increases or decreases with the company's growth, reproduces growth as a general organisational goal since organisations are supposed to provide profits to shareholders or to enable shareholders to make a fortune by selling their shares. Combatting such a growth culture and a concentration of economic power in the hands of a few, post-growth encourages cooperative organisational structures in which "one-member-one-vote" is the rule, creating an egalitarian governance structure. There is also often an upper limit in such organisations on the financial value that any individual can put into an enterprise to prevent it from becoming dependent on that one individual. Dividends are then usually dependent on the face value of the initial and bought value of the share. Moreover, while shares can be withdrawn, they cannot be traded. This arrangement makes organisations "less vulnerable to a one-sided focus on maximizing returns on (potentially spurious) production, growth of production and the scale of its operations" (Johanisova & Wolf, 2012, p. 565). Hence, cooperative organisational structures are a crucial policy idea for achieving communal and collective ownership. The idea of co-housing communities speaks to that same objective. Co-housing communities are neighbourhoods consisting of private and common dwellings, created with the intention of both creating a sense of community and providing private space. Such arrangements are supposed to develop and strengthen non-market relationships and increase collective consumption based on sharing systems (Lietaert, 2010).

The policy ideas discussed above serve post-growth's ambition to collectivise property and the connected notion that commons are local resources that are supposed to be held in collective and communal ownership. Often, specific rights to use the commons are distributed based on need and customs. From this point of view, all have certain rights and duties in relation to the commons, and all need to be involved if profit is made from their use. This entails propagating a communal property rights regime to the detriment of individual private property rights, encouraging local exchanges (Johanisova & Wolf, 2012, p. 567). Furthermore, to encourage local economic exchange, post-growth scholars suggest a stronger regulation of international free trade and capital movement (Daly & Farley, 2011; Jackson, 2009).

Redistribution of income and wealth leads to equal participation in economic processes. If everyone has some degree of control over a certain amount of capital, this empowers the individual, and there is a more scattered power constellation within economic processes (Johanisova & Wolf, 2012, p. 567f.). Generally, the goal of redistribution policies is that individuals should "not be able to capture for themselves values created by nature, by society, or by the work of others. And they should pay a fair price for what they receive from others, including the services provided by the government, and for the costs they impose on others" (Daly & Farley, 2011, p. 442). In this vein, post-growth calls for a minimum income floor and a maximum ceiling (Alexander, 2015; Buch-Hansen & Koch, 2019). A universal basic income can provide a minimum income floor, while the maximum ceiling can be provided by income and wealth caps. A basic income is an unconditional payment to which one is entitled by citizenship. The idea behind an unconditional income is that it provides a certain freedom of choice between paid and unpaid work, improving work–life balance, and contributing to the post-growth's core economy of care work, exchanging time, wisdom, experience, knowledge and skills (Gough, 2017).[5] This would enable people to lead a creative life if they happen not to have formal employment, also acknowledging the importance of non-remunerated care practices. In this thinking, society would evade the urge for ever-increasing employment possibilities, which is closely linked to economic growth. Moreover, the revenues of green taxes could be used to provide a basic income, which would also, in an eco-social policy sense, absorb the potential regressive effect of ecological taxes on low-income households (Andersson, 2010). A related policy proposal to a universal basic income are universal basic services (see Büchs, 2021b). Universal basic services entail the assumption that the satisfaction of basic human needs—such as health care, education, housing, food, mobility and information—should be provided free to everyone (Coote & Percy, 2020).[6]

[5] However, Gough (2017) perceives a basic income as inappropriate for a post-growth strategy, arguing that "one-size-fits-all policies" such as the basic income often fail to succeed and are often characterised by unintended consequences. Nevertheless, other post-growth scholars see a basic income as a key element in a post-growth society (see, e.g., Alexander, 2015; Andersson, 2010; Buch-Hansen, 2014; Büchs & Koch, 2017; Kallis, 2011; Schneider et al., 2010).

[6] While Büchs (2021b) argues that a universal basic income and universal basic services can complement each other well, Coote and Percy (2020, p. 51ff.) argued that they might crowd each other out financially.

A maximum ceiling would be provided by income and wealth caps. The general goal here is to create an egalitarian society. This thinking advocates that a higher urgency should be given to alleviating poverty than to allocating more resources and capital in a world with enough finances to eliminate poverty (Buch-Hansen & Koch, 2019, p. 2). Moreover, proponents of income and wealth caps point to benefits for democracy. Wealthy people often use their wealth and resources to exert political influence, undermining the political equality of ordinary citizens within a democratic polity (Raworth, 2017, p. 172f.; Robeyns, 2017, p. 6ff.). In addition, such caps can be applied to reduce the consumption of natural resources. Tim Jackson (2009) and Kate Raworth (2017) note that inequality is a driving force behind consumption patterns. Accordingly, the wealthy often set the aspirational standard for consumption goods. Decreasing inequality and thus decreasing status competition might also lead to a decrease in ecologically and socially harmful consumption patterns (Demaria et al., 2013, p. 200). Four configurations of income and wealth caps are usually discussed (Buch-Hansen & Koch, 2019). The first and most far-reaching proposal would be a simultaneous cap on income and wealth. This would set limits on what a person can earn in total based on their work, investments and land, and on their wealth assessed on the basis of everything a person has in the way of property with an exchange value minus their debt. The second proposal would be solely a total tax on income, meaning that income would be taxed at 100 per cent if it reached a certain limit. Compared to the first proposal, this is a slower way of fighting inequality, but it would gradually decrease wealth inequality because it makes it relatively costly to maintain luxury goods. The advantage of this proposal is that it is relatively easy to implement technically because income usually takes a monetary form whereas wealth often does not have a fixed monetary value. The third proposal would establish an income cap tied to an income floor, such that a certain income floor is set, and any amount ten times greater is taxed at 100 per cent. Here, the benefit is that once the system was in place, wealthy people would have to work within it to increase the income floor in order to increase their own income which would potentially secure trickle-down effects (Pizzigati, 2004). The fourth proposal would be to establish an income cap on income exceeding an affluence line. This line would serve to indicate a limit beyond which it can be assumed that consumption is no longer practised solely to satisfy an individual's social needs but rather harms the ability of others to satisfy their social needs and the ecology (note the similarity to the idea of sustainable consumption corridors discussed in Sect. 3.3.2) (Buch-Hansen & Koch, 2019). However, the potentially redistributive effect of income and wealth caps might lead to the lower strata creating more demand due to them having more resources available. This could also increase harmful ecological consumption by more social groups (Koch & Fritz, 2014, p. 698). Therefore, from a post-growth perspective, it is crucial to supplement financial caps with recomposing consumption and production policies, such as resource caps, as discussed in PCT schemes, ensuring that ecologically and socially harmful activities decrease on a general level. When thinking about a stable world population, Büchs and Koch (2017, p. 118f.) assume a potentially positive effect from redistribution policies. They state that this might lead to smaller family sizes if they are enacted globally since the elderly would become less dependent on informal family care work. Generally, where population growth is concerned, post-growth emphasises that some decrease in population growth, either in the long run or already in the short term, needs to occur (see, for example, Kerschner, 2010; Schneider et al., 2010). However, policy proposals usually do not go beyond encouraging voluntary population control (Videira et al., 2014), focusing on education in family planning and broad access to contraception connected to equal rights of women (Cosme et al., 2017; Dietz & O'Neill, 2013).

3.5 Transformative Potential of the Eco-Social Policy Perspectives

Section 3.5 offers an overview of the different perspectives and how they relate to each other in terms of existing clashes and congruencies between them. Table 3.1 provides an assessment of the perspectives in terms of their transformative potential, which correlates with the degree to which they enhance public modes of governance (encompassing government intervention as well as cooperative and communal organisational principles that have the potential to transcend the nation-state) and their relationship to economic growth.

Green growth encompasses the least transformative eco-social policy perspectives, with green economy being more moderate than green Keynesianism. Neither version of green growth challenges contemporary policy-making in a profound way, instead they maintain the existing capitalist mode of production at both national and transnational levels. They

Table 3.1 Transformative potential of the various eco-social policy perspectives.

	Degree of public mode of governance	*Relationship towards economic growth*	*Eco-social and transformative potential*
Green growth—green economy	Increasing the scope of private actors while rejecting public modes of governances	Approval of economic growth as crucial to solving environmental and social crises	Least transformative and low awareness of eco-social ideas: "new critical orthodoxy"
Green growth—green Keynesianism	Increasing the scope of government intervention	Approval of economic growth as crucial to solving environmental and social crises	Aware of ecological consequences of social policy, and social consequences of environmental policy; just transition; "new critical orthodoxy"
Recomposing consumption and production	Approval of government intervention and communal modes of possession and consumption	Relatively indifferent towards economic growth; transition between growth-based policy-making and degrowth	Potential to introduce transformative policies; distribution of resource consumption (e.g., consumption and production corridors)
Post-growth	Strong approval of public modes of governance; strong emphasis on (local) community	Abandoning GDP-based growth seen as inevitable in order to stay within planetary boundaries	Most transformative. Envisages a completely different way of organising society; focus on community and care work; decommodification of nature and labour

See also Schulze Waltrup (2025)

align with ecological modernisation that "focus(es) on redirecting and transforming 'free market capitalism' in such a way that it less and less obstructs, and increasingly contributes to, the preservation of society's substance base in a fundamental/structural way" (Mol & Spaargaren, 2000, p. 23).

Green economy represents the most modest eco-social policy perspective. It primarily focuses on the challenge of making economic growth environmentally sustainable by assigning value to nature in monetary and economic terms to incentivise private actors to expand sustainable practices. Its main eco-social policy objective is to address the tension between economic growth and environmental degradation by greening economic growth, thereby allowing traditional and growth-based social policy-making to continue. As a result, social issues are assumed to be dealt with automatically through remedying the tension between economic growth and growth in environmental protection. However, eco-social policy also entails other policy ideas, such as addressing the ecological consequences of specific social policy arrangements or the social consequences of certain environmental policy constellations (Mandelli, 2022). Green economy hardly captures such interrelations.

Green Keynesianism, however, demonstrates a greater understanding of these interrelations. Unlike green economy, it conceptualises the government as a crucial actor in steering green growth developments. It follows social policy objectives by protecting vulnerable parts within a "just transition of ecological modernisation" (Kreinin, 2020) that is generally more aware of the potential regressive effects of environmental policy measures. Green Keynesianism seeks to address the ecological consequences of social policy measures, as well as the social consequences of environmental policy measures. It particularly encourages "investment-oriented eco-social policies" (Mandelli, 2022), which drive economic growth to a certain extent. For instance, it may include labour market policies that support individuals transitioning from brown to green jobs, as emphasised by just transition approaches (Krause et al., 2022). "[P]rotection oriented eco-social policies" (Mandelli, 2022) that are often indifferent to or even oppose economic growth run through all perspectives or, if they explicitly reject economic growth, are part of post-growth.

Nevertheless, like green economy, green Keynesianism values the traditional interconnection between growth, employment and consumer demand. While emphasising social protection and redistributive justice, it still operates within the framework of established patterns of growth and consumption (Birdsall, 2005; Utting, 2013). In general, it and green economy follow the idea of a "new critical orthodoxy" (Brand et al., 2020; see also Sect. 7.3). They reject more radical understandings that explicitly question the existing institutions of contemporary policy-making, such as economic growth, the market or the nation-state. Instead, they treat these institutions as essential components for addressing current crises, provided they are adapted in certain ways. For green Keynesianism, this entails the concept of a "new" steering government that is deemed better equipped than the market to solve pressing issues. However, it still adheres to the capitalist logic of pursuing profitable investments and valorising nature. In that sense, resource extractivism prevails, especially in order to produce new and competitive technologies. Consequently, green Keynesianism is in pursuit of new accumulation strategies that resemble crisis management, ultimately resulting in a form of green nation-state capitalism. This pathway represents an alternative to previous policy models that favoured deregulation and privatisation. However, the "new" entrepreneurial nation-state still relies on economic growth to generate tax revenues, carrying the risk of reproducing the "imperial mode of living". This mode of living normalises unequal exchange relations between the Global North and Global South, rooted in a specific organisation of mass production and mass consumption that sustains a relatively stable set of welfare institutions in the Global North (Brand & Wissen, 2018).

Recomposing consumption and production aligns well with green Keynesianism as both recognise the government's crucial role in intervening in market operations. Therefore, recomposing consumption and production, just like green Keynesianism, stands somewhat in opposition to green economy although it does not necessarily reject the objective of greening economic growth. However, depending on its specific configuration, recomposing consumption and production can also set the stage for post-growth. Like post-growth, it calls for a decrease in consumption and production levels and thus resource throughput. This is also one of its main differences from green Keynesianism. Furthermore, like post-growth, recomposing consumption and production also promotes collective modes of possession. However, recomposing consumption and production can coexist with economic growth, whereas post-growth explicitly rejects the pursuit of increased economic growth, at least in the Global North. In other words, recomposing consumption and production sits somewhere between green Keynesianism and post-growth, potentially continuing to rely on green growth while also potentially paving the way for post-growth (Gough, 2017). This implies that green Keynesianism can serve as a stepping stone towards post-growth, given the connection between it and recomposing consumption and production. Since the pathway towards post-growth would need to start within the existing status quo, green Keynesianism might be a reasonable starting point for a post-growth social-ecological transformation (Custers, 2010).

Post-growth stands out as the most transformative perspective, representing a direct counterperspective to green growth. While green growth aligns with attempts at ecological modernisation to reform capitalism within a process of crisis management, post-growth promotes various elements of a social-ecological transformation that fundamentally reimagines the relationship between society and nature. It advocates communal ownership and the decommodification of nature and labour (Novy, 2022). This also speaks to "de-globalising" certain developments, which exerts pressure to liberalise and "open-up" existing provisioning systems or bring about the expansion of high-carbon infrastructures such as (air)ports, while at the same time still "strengthening cooperation-oriented forms of globalisation" (ibid., p. 75).

This typology facilitates assessment of the OECD and the World Bank from an eco-social policy perspective. Defining the relationship between the different eco-social policy perspectives helps to reconstruct the degree to which the discourses of the two organisations contain eco-social and transformative potential. In other words, this book uses this typology as a basis of comparison for reconstructing the World Bank's and the OECD's eco-social policy discourses (Chaps. 4 and 5).

References

Agrawal, N. (2016). Inequality in India: What's the real story? https://www.weforum.org/agenda/2016/10/inequality-in-india-oxfam-explainer/

Alcott, B. (2013). Should degrowth embrace the job guarantee? *Journal of Cleaner Production, 38*, 56–60. https://doi.org/10.1016/j.jclepro.2011.06.007

Alexander, S. (2015). Basic and maximum income. In G. D'Alisa, F. Demaria, & G. Kallis (Eds.), *Degrowth. A vocabulary for a new era* (pp. 146–148). Routledge.

Andersson, J. O. (2010). Basic income from an ecological perspective. *Basic Income Studies, 4*(2). https://doi.org/10.2202/1932-0183.1180

Bailey, D. (2015). The environmental paradox of the welfare state: The dynamics of sustainability. *New Political Economy, 20*(6), 793–811. https://doi.org/10.1080/13563467.2015.1079169

Bärnthaler, R., & Gough, I. (2023). Provisioning for sufficiency: Envisaging production corridors. *Sustainability: Science, Practice and Policy, 19*(1), 2218690. https://doi.org/10.1080/15487733.2023.2218690

Barry, J. (2012). *The politics of actually existing unsustainability: Human flourishing in a climate-changed, carbon-constrained world*. Oxford University Press.

Beça, P., & Santos, R. (2010). Measuring sustainable welfare: A new approach to the ISEW. *Ecological Economics, 69*(4), 810–819. https://doi.org/10.1016/j.ecolecon.2009.11.031

Beddoe, R., Costanza, R., Farley, J., Garza, E., Kent, J., Kubiszewski, I., et al. (2009). Overcoming systemic roadblocks to sustainability: The evolutionary redesign of worldviews, institutions, and technologies. *Proceedings of the National Academy of Science U S A, 106*(8), 2483–2489. https://doi.org/10.1073/pnas.0812570106

Berr, E. (2015). Sustainable development in a post Keynesian perspective: Why eco-development is relevant to post Keynesian economics. *Journal of Post Keynesian Economics, 37*(3), 459–480. https://doi.org/10.1080/01603477.2015.1000173

Birdsall, N. (2005). *The world is not flat: Inequality and injustice in our global economy*. WIDER Annual Lecture 9, Helsinki: UNU World Institute for Development Economics Research. https://www.cgdev.org/sites/default/files/archive/doc/commentary/speeches/Birdsall_WIDERpaper.pdf

Blackwater, B. (2012). Two cheers for environmental Keynesianism. *Capitalism Nature Socialism, 23*(2), 51–74. https://doi.org/10.1080/10455752.2012.675232

Bowen, A., & Hepburn, C. (2014). Green growth: An assessment. *Oxford Review of Economic Policy, 30*(3), 407–422. https://doi.org/10.1093/oxrep/gru029

Bows, A., & Barrett, J. (2010). Cumulative emission scenarios using a consumption-based approach: A glimmer of hope? *Carbon Management, 1*(1), 161–175. https://doi.org/10.4155/cmt.10.17

Brand, U., & Wissen, M. (2018). *The limits to capitalist nature. Theorizing and overcoming the Imperial mode of living*. Rowman & Littlefield.

Brand, U., Görg, C., & Wissen, M. (2020). Overcoming neoliberal globalization: Social-ecological transformation from a Polanyian perspective and beyond. *Globalizations, 17*(1), 161–176. https://doi.org/10.1080/14747731.2019.1644708

Brunnée, J., & Streck, C. (2013). The UNFCCC as a negotiation forum: Towards common but more differentiated responsibilities. *Climate Policy, 13*(5), 589–607. https://doi.org/10.1080/14693062.2013.822661

Buch-Hansen, H. (2014). Capitalist diversity and degrowth trajectories to steady-state economies. *Ecological Economics, 106*, 167–173. https://doi.org/10.1016/j.ecolecon.2014.07.030

Buch-Hansen, H. (2018). The prerequisites for a degrowth paradigm shift: Insights from critical political economy. *Ecological Economics, 146*, 157–163. https://doi.org/10.1016/j.ecolecon.2017.10.021

Buch-Hansen, H., & Koch, M. (2019). Degrowth through income and wealth caps? *Ecological Economics, 160*, 264–271. https://doi.org/10.1016/j.ecolecon.2019.03.001

Büchs, M. (2021a). Sustainable welfare: Independence between growth and welfare has to go both ways. *Global Social Policy, 21*(2), 323–327. https://doi.org/10.1177/14680181211019153

Büchs, M. (2021b). Sustainable welfare: How do universal basic income and universal basic services compare? *Ecological Economics, 189*, 107152.

Büchs, M., & Koch, M. (2017). *Postgrowth and wellbeing. Challenges to sustainable welfare*. Palgrave Macmillan.

Büchs, M., & Koch, M. (2019). Challenges for the degrowth transition: The debate about wellbeing. *Futures, 105*, 155–165. https://doi.org/10.1016/j.futures.2018.09.002

Büchs, M., Ivanova, D., & Schnepf, S. V. (2021). Fairness, effectiveness, and needs satisfaction: New options for designing climate policies. *Environmental Research Letters, 16*(12). https://doi.org/10.1088/1748-9326/ac2cb1

Cassiers, I. (Ed.). (2015). *Redefining prosperity*. Routledge.

Cato, M. S. (2013). *The paradox of green Keynesianism*. Green House.

Colby, M. E. (1989). *The evolution of paradigms of environmental management in development*. In Strategic Planning and Review Department Working Paper WPS313. World Bank.

Cömert, M. (2019). Revival of Keynesian economics or greening capitalism: "Green Keynesianism". *Sosyoekonomi, 129–144*. https://doi.org/10.17233/sosyoekonomi.2019.04.07

Cooper, G. (2008). *The origin of financial crises: Central banks, credit bubbles and the efficient market fallacy*. Harriman House.

Coote, A., & Goodwin, N. (2010). *The Great Transition: Social justice and the core economy*. NEF Working Paper 1. https://neweconomics.org/uploads/files/82c90c4bb4d6147dc3_1fm6bxppl.pdf

Coote, A., & Percy, A. (2020). *The case for universal basic services*. John Wiley & Sons.

Cosme, I., Santos, R., & O'Neill, D. W. (2017). Assessing the degrowth discourse: A review and analysis of academic degrowth policy proposals. *Journal of Cleaner Production, 149*, 321–334. https://doi.org/10.1016/j.jclepro.2017.02.016

Costanza, R., Cumberland, J. H., Daly, H., Goodland, R., Norgaard, R. B., Kubiszewski, I., & Franco, C. (2015). *An introduction to ecological economics*. CRC Press.

Cotta, B. (2025). The eco-social aspects of the European Green Deal and the Farm to Fork. *Global Social Policy, 25*(1), 112–130. https://doi.org/10.1177/14680181241261068

Custers, P. (2010). The tasks of Keynesianism today: Green new deals as transition towards a zero growth economy? *New Political Science, 32*(2), 173–191. https://doi.org/10.1080/07393141003722032

D'Alisa, G., Deriu, M., & Federico, D. (2015). Care. In G. D'Alisa, D. Federico, & G. Kallis (Eds.), *Degrowth. A vocabulary for a new era* (pp. 63–66). Routledge.

Daly, H. E. (1992). Allocation, distribution, and scale: Towards an economics that is efficient, just, and sustainable. *Ecological Economics, 6*(3), 185–193. https://doi.org/10.1016/0921-8009(92)90024-m

Daly, H. E. (2008). A steady-state economy. In *SDC workshop 'confronting structure'—Achieving economic sustainability.* Sustainable Development Commission.

Daly, H. E., & Farley, J. (2011). *Ecological economics. Principles and applications* (3rd ed.). Island Press.

De Camillis, C., & Goralczyk, M. (2012). Towards stronger measures for sustainable consumption and production policies: Proposal of a new fiscal framework based on a life cycle approach. *The International Journal of Life Cycle Assessment, 18*(1), 263–272. https://doi.org/10.1007/s11367-012-0460-5

Demaria, F., Schneider, F., Sekulova, F., & Martinez-Alier, J. (2013). What is degrowth? From an activist slogan to a social movement. *Environmental Values, 22*(2), 191–215. https://doi.org/10.3197/096327113x13581561725194

Dengler, C., & Lang, M. (2022). Commoning care: Feminist degrowth visions for a socio-ecological transformation. *Feminist Economics, 28*(1), 1–28. https://doi.org/10.1080/13545701.2021.1942511

Devetter, F.-X., & Rousseau, S. (2011). Working hours and sustainable development. *Review of Social Economy, 69*(3), 333–355. https://doi.org/10.1080/00346764.2011.563507

Di Giulio, A., & Fuchs, D. (2014). Sustainable consumption corridors: Concept, objections, and responses. *GAIA—Ecological Perspectives for Science and Society, 23*(3), 184–192. https://doi.org/10.14512/gaia.23.S1.6

Dietz, R., & O'Neill, D. W. (2013). *Enough is enough: Building a sustainable economy in a world of finite resources.* Routledge.

Dittmer, K. (2013). Local currencies for purposive degrowth? A quality check of some proposals for changing money-as-usual. *Journal of Cleaner Production, 54*, 3–13. https://doi.org/10.1016/j.jclepro.2013.03.044

Dittmer, K. (2015). Community currencies. In G. D'Alisa, F. Demaria, & G. Kallis (Eds.), *Degrowth. A vocabulary for a new era* (pp. 149–151). Routledge.

Douthwaite, R. (1999). *The growth illusion. How economic growth has enriched the few, impoverished the many and endangered the planet.* Green Books.

Dresner, S., & Ekins, P. (2006). Economic instruments to improve UK home energy efficiency without negative social impacts. *Fiscal Studies, 27*(1), 47–74. https://doi.org/10.1111/j.1475-5890.2006.00027.x

Druckman, A., & Jackson, T. (2008). Household energy consumption in the UK: A highly geographically and socio-economically disaggregated model. *Energy Policy, 36*(8), 3177–3192. https://doi.org/10.1016/j.enpol.2008.03.021

Dryzek, J. S., Downes, D., Hunold, C., & Schlosberg, D. (2003). *Green states and social movements: Environmentalism in the United States, United Kingdom, Germany, & Norway.* Oxford University Press.

Duit, A., Feindt, P. H., & Meadowcroft, J. (2015). Greening leviathan: The rise of the environmental state? *Environmental Politics, 25*(1), 1–23. https://doi.org/10.1080/09644016.2015.1085218

Easterlin, R. A. (1974). Does economic growth improve the human lot? Some empirical evidence. In P. A. David & M. W. Reder (Eds.), *Nations and households in economic growth* (pp. 89–125). Academic Press.

Elgin, D. (1982). *Voluntary simplicity: An ecological lifestyle that promotes personal and social renewal.* Bantam Books.

Eskelinen, T. (2015). Possibilities and limits of green Keynesianism. In K. Borgnäs, T. Eskelinen, J. Perkiö, & R. Warlenius (Eds.), *The politics of ecosocialism. Transforming welfare* (pp. 101–115). Routledge.

Falkner, R. (2008). *Business power and conflict in international environmental politics.* Palgrave Macmillan.

Falkner, R. (2016). The Paris agreement and the new logic of international climate politics. *International Affairs, 92*(5), 1107–1125. https://doi.org/10.1111/1468-2346.12708

Fanning, A. L., O'Neill, D. W., & Büchs, M. (2020). Provisioning systems for a good life within planetary boundaries. *Global Environmental Change, 64.* https://doi.org/10.1016/j.gloenvcha.2020.102135

Fanning, A. L., O'Neill, D. W., Hickel, J., & Roux, N. (2022). The social shortfall and ecological overshoot of nations. *Nature Sustainability, 5*(1), 26–36. https://doi.org/10.1038/s41893-021-00799-z

Fawcett, T., & Parag, Y. (2010). An introduction to personal carbon trading. *Climate Policy, 10*(4), 329–338. https://doi.org/10.3763/cpol.2010.0649

Fitzpatrick, T. (2011). Challenges for social policy. In T. Fitzpatrick (Ed.), *Understanding the environment and social policy* (pp. 61–90). The Policy Press.

Franzen, A. (2003). Environmental attitudes in international comparison: An analysis of the ISSP surveys 1993 and 2000. *Social Science Quarterly, 84*(2), 297–308. https://doi.org/10.1111/1540-6237.8402005

Fuchs, D. (2007). *Business power in global governance.* Lynne Rienner.

Fuchs, D. (2013). Sustainable consumption. In R. Falkner (Ed.), *The handbook of global climate and environment policy* (pp. 215–230). John Wiley & Sons, Ltd.

Fuchs, D., & Lorek, S. (2001). An inquiry into the impact of globalization on the potential for 'sustainable consumption'. In *Households. ProSus Report 2/01.* ProSus.

Fuchs, D., Sahakian, M., Gumbert, T., Di Giulio, A., Maniates, M., Lorek, S., & Graf, A. (2021). *Consumption corridors. Living a good life within sustainable limits.* Routledge.

Gerber, J.-F., & Raina, R. S. (2018). Post-growth in the global South? Some reflections from India and Bhutan. *Ecological Economics, 150*, 353–358. https://doi.org/10.1016/j.ecolecon.2018.02.020

Goldstein, J., & Tyfield, D. (2017). Green Keynesianism: Bringing the entrepreneurial state back in(to question)? *Science as Culture, 27*(1), 74–97. https://doi.org/10.1080/09505431.2017.1346598

Gough, I. (2016). Welfare states and environmental states: A comparative analysis. *Environmental Politics, 25*(1), 24–47. https://doi.org/10.1080/09644016.2015.1074382

Gough, I. (2017). *Heat, greed and human need. Climate change, capitalism and sustainable wellbeing*. Edward Elgar Publishing.

Gough, I., & Meadowcroft, J. (2011). Decarbonizing the welfare state. In J. S. Dryzek, R. B. Norgaard, & D. Schlosberg (Eds.), *The Oxford handbook of climate change and society* (pp. 490–503). Oxford University Press.

Gough, I., & Therborn, G. (2010). The global future of welfare states. In F. G. Castles, S. Leibfried, J. Lewis, H. Obinger, & C. Pierson (Eds.), *The Oxford handbook of the welfare state* (pp. 703–720). Oxford University Press.

Greening, L. A., Greene, D. L., & Difiglio, C. (2000). Energy efficiency and consumption—The rebound effect—A survey. *Energy Policy, 28*(6–7), 389–401. https://doi.org/10.1016/S0301-4215(00)00021-5

Haas, W., Krausmann, F., Wiedenhofer, D., & Heinz, M. (2015). How circular is the global economy?: An assessment of material flows, waste production, and recycling in the European Union and the world in 2005. *Journal of Industrial Ecology, 19*(5), 765–777. https://doi.org/10.1111/jiec.12244

Hajer, M. A. (1995). *The politics of environmental discourse: Ecological modernization and the policy process*. Oxford University Press.

Hajer, M. A. (2009). Ecological modernisation as cultural politics. In A. P. J. Mol, D. A. Sonnenfeld, & G. Spaargaren (Eds.), *The ecological modernisation reader. Environmental reform in theory and practice* (pp. 80–100). Routledge.

Harris, J. M. (2013). *Green Keynesianism: Beyond standard growth paradigms*. GDAE Working Papers 13–02, GDAE, Medford and Somerville, MA: Tufts University.

Hertwich, E. G. (2005). Consumption and the rebound effect: An industrial ecology perspective. *Journal of Industrial Ecology, 9*(1–2), 85–98. https://doi.org/10.1162/1088198054084635

Hickel, J. (2020). What does degrowth mean? A few points of clarification. *Globalizations*, 1–7. https://doi.org/10.1080/14747731.2020.1812222

Hickel, J., & Kallis, G. (2020). Is green growth possible? *New Political Economy, 25*(4), 469–486. https://doi.org/10.1080/13563467.2019.1598964

Hirsch, F. (2005). *Social limits to growth*. Routledge.

Hirvilammi, T. (2020). The virtuous circle of sustainable welfare as a transformative policy idea. *Sustainability, 12*(1), 391. https://doi.org/10.3390/su12010391

Hirvilammi, T., & Helne, T. (2014). Changing paradigms: A sketch for sustainable wellbeing and ecosocial policy. *Sustainability, 6*(4), 2160–2175. https://doi.org/10.3390/su6042160

Hubacek, K., Baiocchi, G., Feng, K., Muñoz Castillo, R., Sun, L., & Xue, J. (2017). Global carbon inequality. *Energy, Ecology and Environment, 2*(6), 361–369. https://doi.org/10.1007/s40974-017-0072-9

Huber, J. (2004). *New technologies and environmental innovation*. Edward Elgar.

Huber, J., & Robertson, J. (2000). *Creating new money. A monetary reform for the information age*. New Economics Foundation.

Jackson, T. (2005). *Motivating sustainable consumption. A review of evidence on consumer and behavioural change*. Report to the Sustainable Development Research Network.

Jackson, T. (2009). *Prosperity without growth: Economics for a finite planet*. Earthscan.

Jacobs, M. (1991). *The green economy: Environment, sustainable development and the politics of the future*. Pluto Press.

Jacobs, M. (2013). Green growth: Economic theory and political discourse. In R. Falkner (Ed.), *Handbook of global climate and environmental policy* (pp. 197–214). Wiley Blackwell.

Johanisova, N., & Wolf, S. (2012). Economic democracy: A path for the future? *Futures, 44*(6), 562–570. https://doi.org/10.1016/j.futures.2012.03.017

John, P., Smith, G., & Stoker, G. (2009). Nudge nudge, think think: Two strategies for changing civic behaviour. *The Political Quarterly, 80*(3), 361–370. https://doi.org/10.1111/j.1467-923X.2009.02001.x

Kallis, G. (2011). In defence of degrowth. *Ecological Economics, 70*(5), 873–880. https://doi.org/10.1016/j.ecolecon.2010.12.007

Kallis, G. (2017). Radical dematerialization and degrowth. *Philosophical Transactions of the Royal Society A: Mathematical, Physical and Engineering Sciences, 375*(2095). https://doi.org/10.1098/rsta.2016.0383

Kallis, G. (2018). *Degrowth*. Agenda Publishing.

Kallis, G., Kalush, M., O'Flynn, H., Rossiter, J., & Ashford, N. (2013). 'Friday off': Reducing working hours in Europe. *Sustainability, 5*(4), 1545–1567. https://doi.org/10.3390/su5041545

Kallis, G., Kostakis, V., Lange, S., Muraca, B., Paulson, S., & Schmelzer, M. (2018). Research on degrowth. *Annual Review of Environment and Resources, 43*, 291–316. https://doi.org/10.1146/annurev-environ-102017-025941

Kasser, T. (2017). Living both well and sustainably: A review of the literature, with some reflections on future research, interventions and policy. *Philosophical Transactions of the Royal Society A: Mathematical, Physical and Engineering Sciences, 375*(2095). https://doi.org/10.1098/rsta.2016.0369

Kerschner, C. (2010). Economic de-growth vs. steady-state economy. *Journal of Cleaner Production, 18*(6), 544–551. https://doi.org/10.1016/j.jclepro.2009.10.019

Keynes, J. M. (1936). *The general theory of employment, interest and money*. Harcourt, Brace & World, Inc.

King, J. E. (2008). Post-Keynesian economics. In S. Durlauf & L. E. Blume (Eds.), *The new Palgrave dictionary of economics* 2nd ed. (online). Palgrave Macmillan.

Knight, K. W., Schor, J. B., & Jorgenson, A. K. (2017). Wealth inequality and carbon emissions in high-income countries. *Social Currents, 4*(5), 403–412. https://doi.org/10.1177/2329496517704872

Koch, M. (2020). The state in the transformation to a sustainable postgrowth economy. *Environmental Politics, 29*(1), 115–133. https://doi.org/10.1080/09644016.2019.1684738

Koch, M., & Fritz, M. (2014). Building the eco-social state: Do welfare regimes matter? *Journal of Social Policy, 43*(4), 679–703. https://doi.org/10.1017/s004727941400035x

Korpi, W. (1983). *The democratic class struggle*. Routledge and Keagan Paul.

Korpi, W. (1985). Economic growth and the welfare state: Leaky bucket or irrigation system. *European Sociological Review, 1*(2), 97–118.

Krause, D. (2021). Uncovering different degrees of eco-social policy. *Global Social Policy, 21*(2), 332–334. https://doi.org/10.1177/14680181211019165

Krause, D., Stevis, D., Hujo, K., & Morena, E. (2022). Just transitions for a new eco-social contract: Analysing the relations between welfare regimes and transition pathways. *Transfer, 28*(3), 367–382. https://doi.org/10.1177/10242589221127838

Kreinin, H. (2020). *Typologies of 'just transitions': Towards social-ecological transformation*. Ecological Economic Papers, 35., WU Vienna University of Economics and Business, Vienna.

Kubiszewski, I., Costanza, R., Franco, C., Lawn, P., Talberth, J., Jackson, T., & Aylmer, C. (2013). Beyond GDP: Measuring and achieving global genuine progress. *Ecological Economics, 93*, 57–68. https://doi.org/10.1016/j.ecolecon.2013.04.019

Kuhnle, S., & Sander, A. (2010). The emergence of the western welfare state. In F. G. Castles, S. Leibfried, J. Lewis, H. Obinger, & C. Pierson (Eds.), *The Oxford handbook of the welfare state* (pp. 61–80). Oxford University Press.

Lanoie, P., Patry, M., & Lajeunesse, R. (2008). Environmental regulation and productivity: Testing the Porter hypothesis. *Journal of Productivity Analysis, 30*(2), 121–128. https://doi.org/10.1007/s11123-008-0108-4

Latouche, S. (2009). *Farewell to growth*. Polity.

Lebel, L., & Lorek, S. (2008). Enabling sustainable production–consumption systems. *Annual Review of Environment and Resources, 33*(1), 241–275. https://doi.org/10.1146/annurev.environ.33.022007.145734

Lietaert, M. (2010). Cohousing's relevance to degrowth theories. *Journal of Cleaner Production, 18*(6), 576–580. https://doi.org/10.1016/j.jclepro.2009.11.016

Lifset, R. J. (1993). Take it back: Extended producer responsibility as a form of incentive-based environmental policy. *Journal of Resource Management and Technology, 21*(4), 163–175.

Lindellee, J., Alkan Olsson, J., & Koch, M. (2021). Operationalizing sustainable welfare and co-developing eco-social policies by prioritizing human needs. *Global Social Policy, 21*(2), 328–331. https://doi.org/10.1177/14680181211019164

Liobikienė, G., & Dagiliūtė, R. (2016). The relationship between economic and carbon footprint changes in EU: The achievements of the EU sustainable consumption and production policy implementation. *Environmental Science & Policy, 61*, 204–211. https://doi.org/10.1016/j.envsci.2016.04.017

Lipschutz, R. D., & McKendry, C. (2011). In J. S. Dryzek, R. B. Norgaard, & D. Schlosberg (Eds.), *Social movements and global civil society* (pp. 369–383). Oxford University Press.

Lorek, S., & Fuchs, D. (2013). Strong sustainable consumption governance—Precondition for a degrowth path? *Journal of Cleaner Production, 38*, 36–43. https://doi.org/10.1016/j.jclepro.2011.08.008

MacMillan, M., Hoban, T. J., Clifford, W. B., & Brant, M. R. (1997). Social and demographic influences on environmental attitudes. *Journal of Rural Social Sciences, 13*(1), 89–107.

Mandelli, M. (2022). Understanding eco-social policies: A proposed definition and typology. *Transfer: European Review of Labour and Research, 28*(3), 333–348. https://doi.org/10.1177/10242589221125083

Meadowcroft, J. (2005). From welfare state to ecostate. In R. Eckersely & J. Barry (Eds.), *The state and the global ecological crisis* (pp. 3–23). The MIT Press.

Millward-Hopkins, J., & Oswald, Y. (2021). 'Fair' inequality, consumption and climate mitigation. *Environmental Research Letters, 16*(3). https://doi.org/10.1088/1748-9326/abe14f

Mol, A. P. J., & Sonnenfeld, D. A. (2000). Ecological modernisation around the world: An introduction. *Environmental Politics, 9*(1), 1–14. https://doi.org/10.1080/09644010008414510

Mol, A. P. J., & Spaargaren, G. (2000). Ecological modernisation theory in debate: A review. *Environmental Politics, 9*(1), 17–49. https://doi.org/10.1080/09644010008414511

Mol, A. P. J., Sonnenfeld, D. A., & Spaargaren, G. (Eds.). (2009). *The ecological mondernisation reader. Environmental reform in theory and practice*. Routledge.

Muraca, B. (2012). Towards a fair degrowth society: Justice and the right to a 'good life' beyond growth. *Futures, 44*(6), 535–545. https://doi.org/10.1016/j.futures.2012.03.014

Newell, P., & Paterson, M. (2010). *Climate capitalism. Global warming and the transformation of the global economy*. Cambridge University Press.

Novy, A. (2022). The political trilemma of contemporary social-ecological transformation—Lessons from Karl Polanyi's *The Great Transformation*. *Globalizations, 19*(1), 59–80. https://doi.org/10.1080/14747731.2020.1850073

Nussbaum, M. (2003). Capabilities as fundamental entitlements: Sen and social justice. *Feminist Economics, 9*(2–3), 33–59. https://doi.org/10.1080/1354570022000077926

Nyfors, T., Linnanen, L., Nissinen, A., Seppälä, J., Saarinen, M., Regina, K., et al. (2020). Ecological sufficiency in climate policy: Towards policies for recomposing consumption. *Futura, 3*.

O'Neill, K. (2009). *The environment and international relations*. Cambridge University Press.

O'Neill, D. W. (2015). The proximity of nations to a socially sustainable steady-state economy. *Journal of Cleaner Production, 108*, 1213–1231. https://doi.org/10.1016/j.jclepro.2015.07.116

O'Neill, D. W., Fanning, A. L., Lamb, W. F., & Steinberger, J. K. (2018). A good life for all within planetary boundaries. *Nature Sustainability, 1*(2), 88–95. https://doi.org/10.1038/s41893-018-0021-4

Oswald, Y., Owen, A., & Steinberger, J. K. (2020). Large inequality in international and intranational energy footprints between income groups and across consumption categories. *Nature Energy, 5*(3), 231–239. https://doi.org/10.1038/s41560-020-0579-8

Pitkänen, A., Von Wright, T., Kaseva, J., & Kahiluoto, H. (2022). Distributional fairness of personal carbon trading. *Ecological Economics, 201*, 107587. https://doi.org/10.1016/j.ecolecon.2022.107587

Pizzigati, S. (2004). *Greed and good: Understanding and overcoming the inequality that limits our lives*. The Apex Press.

Porter, M., & van der Linde, C. (1995). Toward a new conception of the environment—Competitiveness relationship. *The Journal of Economic Perspectives, 9*(4), 97–118. https://doi.org/10.1257/jep.9.4.97

Raftery, A. E., Zimmer, A., Frierson, D. M. W., Startz, R., & Liu, P. (2017). Less than 2 degrees C warming by 2100 unlikely. *Nature Climate Change, 7*, 637–641. https://doi.org/10.1038/nclimate3352

Rao, S., Pachauri, S., Dentener, F., Kinney, P., Klimont, Z., Riahi, K., & Schoepp, W. (2013). Better air for better health: Forging synergies in policies for energy access, climate change and air pollution. *Global Environmental Change, 23*(5), 1122–1130. https://doi.org/10.1016/j.gloenvcha.2013.05.003

Raworth, K. (2017). *Doughnut economics*. Random House Business Books.

Reyes-Garcia, V., Babigumira, R., Pyhala, A., Wunder, S., Zorondo-Rodriguez, F., & Angelsen, A. (2016). Subjective wellbeing and income: Empirical patterns in the rural developing world. *Journal of Happiness Studies, 17*(2), 773–791. https://doi.org/10.1007/s10902-014-9608-2

Richardson, K., Steffen, W., Lucht, W., Bendtsen, J., Cornell, S. E., Donges, J. F., et al. (2023). Earth beyond six of nine planetary boundaries. *Science. Advances, 9*(37). https://doi.org/10.1126/sciadv.adh2458

Robeyns, I. (2017). Having too much. In J. Knight & M. Schwartzberg (Eds.), *Wealth, nomos LVIII* (pp. 1–44). New York University Press.

Ryan, R. M., Huta, V., & Deci, E. L. (2008). Living well: A self-determination theory perspective on eudaimonia. *Journal of Happiness Studies, 9*(1), 139–170. https://doi.org/10.1007/s10902-006-9023-4

Sakamoto, T. (2020). Social investment policy, economic growth, and welfare states: Channels of pro-growth effects of policy. *Social Forces, 99*(2), 590–615. https://doi.org/10.1093/sf/soz178

Sanne, C. (2002). Willing consumers—Or locked-in? Policies for a sustainable consumption. *Ecological Economics, 42*(1–2), 273–287. https://doi.org/10.1016/S0921-8009(02)00086-1

Schneider, F., Kallis, G., & Martinez-Alier, J. (2010). Crisis or opportunity? Economic degrowth for social equity and ecological sustainability. Introduction to this special issue. *Journal of Cleaner Production, 18*(6), 511–518. https://doi.org/10.1016/j.jclepro.2010.01.014

Schor, J. B. (2004). *Born to buy: The commercialized child and the new consumer culture*. Scribner.

Schor, J. B. (2015). Work sharing. In G. D'Alisa, F. Demaria, & G. Kallis (Eds.), *Degrowth. A vocabulary for a new era* (pp. 195–198). Routledge.

Schröder, E., & Storm, S. (2020). Economic growth and carbon emissions: The road to 'hothouse earth' is paved with good intentions. *International Journal of Political Economy, 49*(2), 153–173. https://doi.org/10.1080/08911916.2020.1778866

Schulze Waltrup, R. (2025). An eco-social policy typology: From system reproduction to transformation. *Global Social Policy, 25*(1), 17–35. https://doi.org/10.1177/14680181231205777

Schumpeter, J. A. (1961 [1943]). *Capitalism, socialism and democracy*. George Allen and Unwin.

Sen, A. (1999). *Development as freedom*. Oxford University Press.

Seyfang, G., & Paavola, J. (2008). Inequality and sustainable consumption: Bridging the gaps. *Local Environment, 13*(8), 669–684. https://doi.org/10.1080/13549830802475559

Shove, E. (2014). Putting practice into policy: Reconfiguring questions of consumption and climate change. *Contemporary Social Science, 9*(4), 415–429. https://doi.org/10.1080/21582041.2012.692484

Soper, K. (2008). Alternative hedonism, cultural theory and the role of aesthetic revisioning. *Cultural Studies, 22*(5), 567–587. https://doi.org/10.1080/09502380802245829

Spangenberg, J. H. (2010). The growth discourse, growth policy and sustainable development: Two thought experiments. *Journal of Cleaner Production, 18*(6), 561–566. https://doi.org/10.1016/j.jclepro.2009.07.007

Stern, P. C., & Aronson, E. (1984). *Energy use: The human dimension*. Freeman.

Stiglitz, J., Sen, A., & Fitoussi, J.-P. (2010). *Mis-measuring our lives. Why GDP doesn't add up*. The New Press.

Swenson, P. (2002). *Capitalists against markets: The making of labor markets and welfare states in the United States and Sweden*. Oxford University Press.

Thaler, R. H., & Sunstein, C. R. (2008). *Nudge: Improving decisions about health, wealth and happiness*. Yale University Press.

Thomas, A. (2020). *Reassessing the regressivity of the VAT*. OECD Taxation Working Papers, (49), 0_1–51.

Thomas, A., & Doerflinger, N. (2020). Trade union strategies on climate change mitigation: Between opposition, hedging and support. *European Journal of Industrial Relations, 26*(4), 383–399. https://doi.org/10.1177/0959680120951700

Tienhaara, K. (2018). *Green Keynesianism and the global financial crisis*. Routledge.

Tienhaara, K., & Robinson, J. (2022). Introduction. In K. Tienhaara & J. Robinson (Eds.), *Routledge handbook on green new deal*. Routledge.

Tukker, A., Diaz Lopez, F. J., van de Lindt, M., Mont, O., Lorek, S., Spaneberg, J., & Giljum, S. (2009). *Sustainable consumption policies effectiveness evaluation (Scope2)*. Report to European Commission within FP6.

Unti, B. J. (2015). Job guarantee. In G. D'Alisa, F. Demaria, & G. Kallis (Eds.), *Degrowth. A vocabulary for a new era* (pp. 172–174). Routledge.

Utting, P. (2013). Pathways to sustainability in a crisis-ridden world. In R. Genevey, R. K. Pachauri, & L. Tubiana (Eds.), *Reducing inequalities: A sustainable development challenge* (pp. 175–190). TERI Press.

Vadén, T., Lähde, V., Majava, A., Järvensivu, P., Toivanen, T., & Eronen, J. T. (2020). Raising the bar: On the type, size and timeline of a 'successful' decoupling. *Environmental Politics, 30*(3), 462–476. https://doi.org/10.1080/09644016.2020.1783951

Vadén, T., Lahde, V., Majava, A., Jarvensivu, P., Toivanen, T., Hakala, E., & Eronen, J. T. (2020). Decoupling for ecological sustainability: A categorisation and review of research literature. *Environmental Science & Policy, 112*, 236–244. https://doi.org/10.1016/j.envsci.2020.06.016

Videira, N., Schneider, F., Sekulova, F., & Kallis, G. (2014). Improving understanding on degrowth pathways: An exploratory study using collaborative causal models. *Futures, 55*, 58–77. https://doi.org/10.1016/j.futures.2013.11.001

Wahlen, S., Heiskanen, E., & Aalto, K. (2012). Endorsing sustainable food consumption: Prospects from public catering. *Journal of Consumer Policy, 35*(1), 7–21. https://doi.org/10.1007/s10603-011-9183-4

Ward, J. D., Sutton, P. C., Werner, A. D., Costanza, R., Mohr, S. H., & Simmons, C. T. (2016). Is decoupling GDP growth from environmental impact possible? *PLoS One, 11*(10), e0164733. https://doi.org/10.1371/journal.pone.0164733

Waring, M. (1988). *If women counted: A new feminist economics*. Harper & Row.

Wier, M., Birr-Pedersen, K., Jacobsen, H. K., & Klok, J. (2005). Are CO2 taxes regressive? Evidence from the Danish experience. *Ecological Economics, 52*(2), 239–251. https://doi.org/10.1016/j.ecolecon.2004.08.005

Wilkinson, R., & Pickett, K. (2009). *The spirit level. Why more equal societies almost always do better*. Allen Lane.

Zenghelis, D. (2012). *A strategy for restoring confidence and economic growth through green investment and innovation*. Retrieved from Grantham Research Institute on Climate Change and the Environment Centre for Climate Change Economics and Policy. (https://www.greenpolicyplatform.org/sites/default/files/downloads/resource/A_Strategy_for_Restoring_Confidence_and_Economic_Growth_CCCEP.pdf)

CHAPTER 4

The OECD's Eco-Social Policy Discourse

This chapter provides a reconstruction of the OECD's eco-social policy discourse. While this introductory section describes the OECD more generally, the two subsequent sections provide a brief bird's-eye view of the OECD's global social and environmental policy, which serves as a contextualisation of the organisation's eco-social policy to reveal a number of (dis)continuities. The review of the OECD's involvement in social and environmental policy is also contextualised within the framework of Polanyi's double movement and indicates that the double movement is present in overarching policy dynamics, whether the policy be environmental, social or eco-social (Sect. 7.1). The review also provides examples of the social and environmental policy issues that are discussed within the OECD more generally, but that are also somewhat detached from an integrated eco-social policy perspective. The third section provides an empirical reconstruction of the OECD's eco-social policy discourse, for which the developed eco-social policy typology (Chap. 3) serves as the analytical framework.

Within the global governance system, the OECD figures as a soft-power organisation. It does not have any financial or legal authority to direct governments to implement certain policies; instead, it uses its expertise and ideas to convince governments to adopt specific policies and aims to diffuse certain policy "best-practices" (Mahon & McBride, 2009). It largely relies on the surveillance and monitoring of member states'

R. Schulze Waltrup, *Eco-Social Policy in the OECD and the World Bank*, https://doi.org/10.1007/978-3-031-95429-0_4

policies, offering commentary and providing a platform for peer evaluation and expert discussions on the best approaches to policy problems (Mahon, 2015). This includes the creation of a shared policy culture and common narratives (Woodward, 2009).

In terms of organisational structure, the OECD comprises various directorates (OECD staff that coordinate committees) connected with its committees (where member states are represented and oversee the directorate's work). Among these units, the Environment Directorate is most closely linked to environmental policy, while the Directorate for Employment, Labour, and Social Affairs is most closely connected with social policy. Generally, the OECD's organisational structure consists of policy silos similar to those found in nation-state ministries with some initiatives that try to overcome this siloed policy-making. The OECD provides policy guidance largely to the high-income countries that are its members. The next two sections reconstruct the OECD with respect to both its global social and its global environmental policy.

4.1 The OECD in Global Social Policy

To contextualise the discussion of the OECD's eco-social policy discourse, this section provides a review of social policy developments within the OECD. Generally, from the mid-1990s on, the OECD steadily emancipated itself from a neoliberal approach. Indeed, its take on different social policy fields sometimes goes quite a long way towards abandoning the strong welfare state retrenchment approach of the 1980s and 1990s. However, other elements have remained remarkably close to that neoliberal agenda. Identifying this trend, the book suggests that the double movement dynamic is clearly present in the OECD's global social policy-making. By way of illustration, this section provides a broad summary of social policy developments within the OECD while also focusing on debates concerning specific social policy fields, including the well-studied cases of labour market policy, education policy, health policy and youth unemployment.

The OECD, like the World Bank (Chap. 5), increasingly acknowledges a productive role for social policy in terms of economic development, which, in one way or the other, increases human capital and therefore would be important for economic growth. This productive role of social policy is well exemplified by the social investment perspective (Deeming & Smyth, 2018; Jenson, 2010). While both the OECD and the World Bank (see Sect. 5.1) apply a social investment perspective in their policy prescriptions, they emphasise different elements. In the field of child development, for example, the OECD lays a particular emphasis on universality, gender equality and child development, in particular favouring policy constellations of the Nordic welfare system that includes scepticism concerning missing shared parental leave policies and long maternal-focused childcare leave. Nevertheless, the OECD also supports some elements of a New Public Management approach that supports the private provision of childcare (Mahon, 2010).

The OECD pushed social investment as a modernisation of social policy to facilitate rediscovering the value of social policy for economic growth. Accordingly, it emphasises the importance of supporting individuals to be integrated into the labour market. This comes with a conception that goes beyond having only a safety net that prevents the poorest from severe destitution in favour of one that focuses on enabling them to participate in the labour market. Therefore, social investment is not solely an "anti-poverty" approach, instead it integrates the middle class as beneficiaries of social policy. It revolves around a positive attitude to government intervention in regard to education, childhood development and anti-poverty. Generally, it argues that social policy is an investment in economic growth (Jenson, 2010, p. 72).

The culmination of conceiving social policy along the lines of social investment and liberal approaches that are more socially inclusive was "New Orientations for Social Policy" (OECD, 1994; see Mahon, 2009). "New Orientations for Social Policy" indicates scepticism towards a neoliberal approach and frames traditional economic approaches and social policy as mutually beneficial. "Non-inflationary growth of output and jobs, and political and social stability are enhanced by the role of social expenditure as investments in society" (OECD, 1994, p. 12). The report states that there needs to be a reasonable balance between public and private provision of services, and that both modes of provisioning have their strengths and weaknesses, which would also be why the two in combination can synergise (Mahon, 2009; OECD, 1994). This signalled a shift from the perception within the OECD of the 1980s that "[s]ocial policy in many countries creates obstacles to growth" (OECD, 1981 quoted in Deacon, 2007). However, in some social policy fields, the OECD does not advocate as a coherent actor, as is also the case with eco-social policy (Sect. 4.3), which generally supports the usefulness of the concept of the double movement to make sense of the policies that the OECD promotes.

For example, the Economics Department and the Directorate for Education, Employment, Labour and Social Affairs had opposing views on labour market policy (McBride & Williams, 2001). The Economics Department promoted a job strategy that advanced the social policy-globalisation narrative of a "race to the bottom", which assumes that increasing economic integration and the associated free flow of capital require deregulated labour markets and low social provisions as capital will simply move where it can best unfold and grow. This narrative comes with policy prescriptions that focus on non-inflationary growth, accelerate the diffusion of technology and knowledge, promote flexibility in terms of working time, wages and labour costs, incentivise entrepreneurship and promote employment in the private sector as well as education programmes to equip individuals with skills for the labour market. In contrast, the Directorate for Education, Employment, Labour and Social Affairs advanced precisely the opposite narrative that social protection and employment protection would have hardly any effect on employment rates, instead leading to an increase in the number of stable jobs. However, the Economics Department had more resources available to successfully diffuse its labour market policy model (McBride & Williams, 2001).

In the field of education policy, the OECD advocates investments in human capital and economically related learning practices. These are promoted by large-scale assessments of member countries. Here, the OECD's PISA (Programme for International Student Assessment) ranking works as a governance tool for creating assessments of how states perform in education. Generally, the OECD perceives education as an investment in human capital, with public authorities bearing responsibility for its provision. However, the OECD primarily emphasises the economic utility of education, often overlooking other important dimensions (Niemann & Martens, 2018, p. 272), suggesting to some extent that public education should predominantly serve the economic sphere. The value of education is considered to be dependent on how it can be utilised in technological progress or in applied research, and the main goal of education is to increase the economic competitiveness of member-states (ibid.).

In terms of health policy, OECD has historically had quite an economic perspective. Even in the 1990s, the organisation struggled to adopt a more comprehensive and holistic view, one that recognises health as a socio-economic concept vital for both economic development and social cohesion. However, as Kaasch (2015) notes, the OECD was rather characterised by not necessarily promoting one particular healthcare system model but usually discussing the different advantages and drawbacks of the various systems that it could identify among member-states. Also, there seems to have been some cautiousness about prescribing specific policy recommendations. Nevertheless, the OECD's policy priority moved towards a more balanced approach of not only considering the economic and financial dimensions of healthcare systems but also its contribution to social cohesion (Kaasch, 2015).

In the field of youth unemployment, the OECD (like the World Bank) focused on human capital and deficiencies in the labour market supply with an understanding that problematised individuals as having the wrong skills and knowledge. With the global financial crisis, the OECD in particular shifted towards a more social-democratic approach that recognised that youth unemployment can also be due to structural issues that impede individuals from taking up these skills and gave additional acknowledgement to expanding social safety nets of unemployment benefits. However, this shift was rather short-lived. As high unemployment rates became normal, the OECD shifted back to focusing on individual skill deficits and stating that comprehensive social safety nets would create preventable fiscal difficulties (Fergusson, 2021).

Another more recent theme within the OECD's social policy discourse, somewhat connected to the idea of social investment (Smyth & Deeming, 2018), is that of inclusive growth. This came along with an inclusive growth framework established in 2012 with strong support from the Secretary-General's office. At the same time, and before that, there was also an emergent focus on inequality and equity within the Directorate of Employment, Labour and Social Affairs (OECD, 2008, 2011d, 2015e). Generally, the framework entails a strong appreciation of social protection and equity, a new focus on highlighting job quality and not necessarily job quantity, as well as a more assertive attitude to the distributional impact of policies and applying multi-dimensional perspectives on well-being (Mahon, 2019).

In sum, this brief review suggests that the double movement dynamic of a struggle between a more neoliberal perspective favouring welfare-state retrenchment and the expansion of providing private services (the first movement), and a countermovement favouring social protection, government intervention and decommodification has shaped social policy development in the OECD. Recently, there has been a shift towards social protection approaches; nevertheless, there is a tendency to conceptualise social policy as being in service to the market and economic growth.

4.2 The OECD in Global Environmental Policy

Following on from the previous section's review of the OECD in global social policy, this section discusses the OECD's involvement in environmental policy. It too suggests that the dynamic of the double movement is at work, here driving the OECD's environmental policy-making while historically favouring the movement of economic liberalisation and (green) self-regulating markets.

Compared to the World Bank and other IOs, the OECD started relatively early to work on environmental policy issues. The Organisation for European Economic Co-operation (the predecessor to the OECD) already contained a commission that worked on environmental policy issues, specifically pollution control, even though that was not the organisation's mandate. In 1961, with the establishment of the OECD, this became the Committee for Scientific Research, adding noise and pesticides and co-operation with other directorates. In 1970, the Committee for Scientific Research became the Environment Committee, which means that it was established before most governments had environmental ministries (and before the United Nations Environment Programme (UNEP) was established in 1972). Generally, by comparison with other international bodies, the OECD started to work relatively early on ecological issues, such as acid rain, ozone depletion, waste and climate change before they were considered important topics within international and national policy debates (Borowy, 2017a).

The OECD influences environmental policy debates in multiple ways. Generally, it provides a forum for high-level government staff to exchange views. Moreover, it provides policy prescriptions and works towards the harmonisation of environmental regulations in member-countries with the general objective of not creating any trade distortions. It sets international standards, for example, for air pollution and chemicals and provides "products" such as technical analysis for chemical waste management, air pollution and indicators of environmental trends in countries. Furthermore, it reviews the ecological practices of member countries to promote accountability and responsibility. The OECD also promotes a synergetic view of ecological and economic issues. In particular, the Environment Committee and its Directorate reject unconditional growth but instead argue for a perspective that takes account of growth that should have an explicit ecological dimension. This talk was already prominent in the 1970s (Long, 2000, p. 121), which suggests that ideas of green growth were already present at least in the Environment Committee and Directorate before the OECD published its green growth strategy prominently in 2011.

The core of the OECD's environmental policy discourse is its Environment Directorate and Committee. Generally, the Directorate works towards the integration of ecological issues into other issue areas, such as trade or energy policy and aims to reduce the negative ecological impacts of "chemicals, energy production and consumption, transport, waste, and agriculture, as well as the protection of biodiversity and the global climate" (Busch, 2009, p. 76). It also coordinates exchange between the Environment Committee and other Committees, and prepares council decisions and the adoption of legal instruments (Busch, 2009, p. 77). Moreover, the Directorate and the Committee have worked to change the perception that ecological protection practices would be bad for economic growth and pushed the institutionalisation of the perspective that ecological protection is crucial to drive sustainable innovation, competition and to create economic growth (ibid., p. 78).

Since its inception, the Environment Directorate has particularly promoted market-based measures, which, for example, in the field of taxes, take account of the social and environmental costs of certain practices and products. Another closely connected policy tool is cost-benefit analysis that investigates the value of environmental protection, biodiversity (loss) and its connected human health risks and the effect of air pollution in monetary terms (Barde, 2007, p. 38). Such market-based policy tools go hand in hand with green economy's (Sect. 3.2.1) ambitions to increase the scope of private actors in promoting ecological protection practices.

This aligns with the fact that, two years after its inception, the Environment Committee released the *Guiding Principles Concerning the International Economic Aspects of Environmental Policies*, which introduced the prominent polluter pays principle (PPP), a key concept in environmental policy debates. When the PPP was introduced by the OECD in the early 1970s, it had a rather limited meaning, being based on the idea that pollution abatement should not be subsidised by the government as this would distort free-trade practices, but instead that the polluters themselves should pay. However, there was at least an implicit meaning that ecological costs should be integrated into the practices of polluters. This latter meaning dominates nowadays (Bernstein, 2001, p. 51). Generally, "[t]he introduction of the PPP marked the start of a trend to incorporate environmental costs into production, markets, and accounting practices, rather than favour what economists label command-and-control regulation to combat environmental damage" (ibid.). Therefore, the PPP is quite in line with the practices of a (green) self-regulating market. However, in the discussions around establishing PPP, there was also remarkable scepticism about solely applying market mechanisms, suggesting that the dynamic of the double movement also characterised negotiations around the PPP.

> It is already common knowledge that economics, the science of managing scarce resources so as to maximise human welfare, is still far from embracing all the factors which contribute to human welfare. Among these factors the provision of a healthy and balanced environment has been widely neglected, mainly because market mechanisms are not capable of taking account of them. (Note by the Secretariat, The "Polluter-Pays" Principle and Vehicle Pollution and Noise Control, 10 November 1971, 4. quoted in Borowy, 2017b, p. 323)

Another OECD measure to potentially influence member-countries' policy-making is to carry out environmental performance reviews (EPRs), which review various policy practices in different environmental fields, such as water management, biodiversity protection and international cooperation. They assess countries based on the country's own policy objectives, international commitments, commonly agreed OECD policy principles and the cost-efficiency of policies (Lehtonen, 2007). Generally, these reviews facilitate actors (in particular, civil servants) pushing for market-based and economics-inspired thinking in environmental policy debates. However, within the OECD's strategic operations, the EPRs do not have a particularly high standing in the sense that other activities and operations could be directed from the EPR, but instead the "causal chain goes in the opposite direction; the overall OECD policy agenda influences the choice of themes and concepts in EPRs" (Lehtonen, 2007, p. 23). In that sense, the EPR reflects a policy agenda that is characterised by an integration of economic issues into the environmental policy agenda. However, the integration of ecological issues into economic reviews is not always successful. In the conceptualisation of economic reviews, notions of sustainable development have been perceived as an agenda from the "outside" (even though they come from inside the OECD). Generally, the organisation is dominated by economic networks, and the organisational sectoral approach undermines the emergence and discussion of alternative views. Hence, implicitly the economic networks influence the EPRs but the environmental network, or the Environment Directorate, does not necessarily influence economic thinking to any large extent. This is due to the privileged and dominant position of economic networks within the organisation (ibid., p. 31).

In sum, traditionally the OECD's environmental policy discourse has largely promoted market-based policy ideas and ideas to incorporate an economic perspective into environmental policy issues, but not necessarily the other way around. Hence, the movement to establish a self-regulating market, at least historically, seems to be dominant within the OECD's environmental policy discourse which has largely been about "getting the price right" and incorporating ecological costs into economic decision-making, reiterating the commodification of nature.

4.3 The OECD's Eco-Social Policy Discourse: Highlighting Inequality and Public Transitions

This section provides an empirical reconstruction of the OECD's eco-social policy discourse based on the eco-social policy typology developed in Chap. 3. This summary and categorisation of various OECD policy elements from an eco-social perspective are developed further in theoretical terms in Chap. 7, in which I present four analytical arguments that deepen our understanding of the global policy and governance dynamics that underpin the OECD's (and the World Bank's) eco-social policy discourse(s). Following the different categories of the eco-social policy typology, this section reconstructs the OECD's policy discourse in terms of green growth, green economy, green Keynesianism, recomposing consumption and production and post-growth, while interconnections and tensions between environmental and social policy are discussed as part of the green Keynesian understanding.

4.3.1 The OECD and Green Growth

This current subsection reconstructs the OECD's green growth understanding. As discussed in the eco-social policy typology (Chap. 3), green growth is an umbrella perspective covering green economy and green Keynesianism (two conflicting eco-social policy perspectives, both following green growth) that aims to solve the conflicts between economic growth and social policy and economic growth and environmental policy by dint of making growth green (Sect. 3.1). Here I provide a rather broad reconstruction of green growth, including the more general growth understanding of the OECD and how this is also connected with social policy issues. The green economy and green Keynesianism are discussed in more detail below.

The OECD established its green growth strategy at the OECD Ministerial Council meeting on 25–26 May 2011 as a flagship initiative within the OECD, and much of its subsequent work refers to the notion of green growth. It was established in the wake of the global financial crisis and calls for economies to be changed in such a way that they no longer damage the ecology but instead are a fundamental part of ecological protection practices. As former Secretary-General Gurría put it in his introduction to the green growth strategy, it is about embarking away from "business as usual" (OECD, 2011a). The strategy was mainly based on work of the Economics Department, Environment Directorate, Statistics Directorate and Directorate for Science, Technology and Industry, so that social affairs were not organisationally represented in it. Leadership was carried out by the Economics Department. Also, the OECD later published revisions and stock-takings of the green growth strategy published in 2011 (OECD, 2015a).

As part of the green growth strategy, the OECD established a series of publications, including Green Growth Papers and Green Growth Studies in which various issues of green growth are discussed. This is also the case with *The Inequalities–Environment Nexus: Towards a People-Centred Green Transition*, which, on the organisational level, is an explicit outcome document from a merger of the inclusive and green growth stream and entails crucial components in particular for the discussion of interconnections and tensions within the OECD's green Keynesianism understanding (see Sect. 4.3.3). Many elements of the OECD's green growth perspective stem from its initial strategy document *Towards Green Growth* (OECD, 2011a). However, a substantial number of elements of the OECD's green growth discourse presented here are also based on the document *Investing in Climate, Investing in Growth* (OECD, 2017a), which was coordinated under the leadership of the Environment Directorate and produced in the context of the G20 presidency of Germany. Other directorates were also involved, including the Economics Department, the Directorate for Financial and Enterprise Affairs, the Development Cooperation Directorate, the Trade and Agriculture

Directorate, the Centre for Tax Policy and Administration, the International Transport Forum and the Directorate for Public Governance. Other important elements of the OECD's green growth understanding stem from the New Approaches to Economic Challenges (NAEC) initiative. The other OECD documents that are referred to in this subsection are all part of different publication series. Overall, they are usually produced by environmental policy working groups within the OECD.

The OECD promotes various themes within its green growth understanding. Based on a general problematisation of economic growth in particular with regard to the ecology, but also in relation to social policy issues, the OECD also connects green growth measures explicitly with social policy objectives. General typical green growth themes (Sect. 3.2) are discussed as well. This, for example, entails allocating ecological resources and services an economic value in order to incorporate ecological thresholds into economic decision-making. Also, ideas of "increasing efficiency" and "increasing productivity" to safeguard the ecology are prominent. Moreover, the OECD promotes the concept of a "knowledge economy" that disregards reliance on high-carbon practices but should be based on low-carbon knowledge practices. Here, the OECD emphasises the importance of research, development and innovation while advancing "first-mover advantages". The OECD also discusses the potentially beneficial relationship of green growth to social policy while also referring to infrastructure as crucial to enabling green growth. All these themes are further explored below.

More generally, the OECD discusses economic growth critically both in relationship to environmental and social policy issues. Accordingly, the OECD acknowledges the detrimental historical impact of economic growth on the ecology: "severe environmental degradation has forced a recognition that today's patterns of economic growth are undermining our capacity to maintain current standards of living" (OECD, 2020a, p. 15). As part of the green growth strategy, it states that "risks to growth and development continue to rise as traditional growth models negatively affect the physical environment that ultimately underpins human well-being" (OECD, 2015a, p. 20). With regard to social policy objectives, the OECD applies the traditional conception (also acknowledging exceptions) that economic growth would have provided an increase in well-being and welfare but states that the environmental cost that is historically related to such an increase endangers future well-being and thus future social policy objectives:

> Economic growth over the last two centuries has led to staggering increases in wealth and well-being for much—but not all—of the world's population. To continue to improve well-being over a longer time horizon, the sources of growth need to be sustainable economically, socially and environmentally. To date, growth has exploited natural capital to meet the demands of rising populations, using technology largely based on abundant fossil fuels. Those fuels have been cheap because little account has been taken of their social and environmental costs. (OECD, 2017a, 18)

Furthermore, the OECD criticises the way growth is often considered to be a policy objective in itself and rejects the existing "silo approach in policy making [which] tended to underplay issues such as inequality and environmental harm, [...] often treated as possible secondary effects to be addressed at a later stage" (OECD, 2015b, p. 9).

NAEC also expresses scepticism concerning the unconditional policy focus on economic growth: "Economic growth continues to generate the benefits of higher national income. But at the same time, the dominant patterns of growth in OECD countries over recent decades have also generated significant harms" (OECD, 2020a, p. 15)—for example, because economic growth often comes with rising inequality (OECD, 2008, 2015e, 2020a). Moreover, the OECD acknowledges that increases in economic growth would be indifferent to more comprehensive categories of the well-being of individuals such as "the security and satisfaction they experience in work; their physical and mental health, social networks and personal and family relationships, and from social goods such as the levels of crime and trust in society, and the quality of public services such as health and education" (OECD, 2020a, p. 15). Indeed, GDP could, in fact, be harmful to those categories of well-being depending on the way that GDP is created (OECD, 2020a, p. 15). Elsewhere, as part of the *Green Growth and Sustainable Development Forum. Addressing the social implications of green growth*, the OECD emphasises in a similar thought that

> [t]he fact that social inequalities reinforce environmental problems means that the growth model inherited from the twentieth century has proven unsustainable not only because of its responsibility in excessive emissions of greenhouse gases disturbing climate, but also because of its contribution to increasing inequality, instability and economic crises, the most recent one being compared to the great depression of the 1930s. (OECD, 2014, p. 10)

In order to connect green growth and social policy, the OECD promotes "green inclusive growth" (OECD, 2021a). This speaks to "investment in human capital, through active labour market policies, well-targeted income support measures, and upgrading skills to facilitate labour reallocation" (ibid., p. 4)—notions that are also well captured under the theme of a just transition (OECD, 2017a). In addition, the OECD states that government should ensure that productivity increases "are broadly shared through higher wages and better employment opportunities" (OECD, 2018a, p. 25). Moreover, the OECD highlights social policy effects of energy efficiency measures. Such measures might have a positive impact on creating jobs since they are "often generally labour-intensive" (OECD, 2020c, p. 10). These notions already represent crucial elements of the OECD's understanding of green Keynesianism (for more on this, see Sect. 4.3.3).

In general, the OECD promotes various themes that are part of green growth as laid out in this book's typology (Chap. 3). Accordingly, the OECD notes that environmental goods are crucial for economic practices and, in order to prevent an unsustainable use of resources, their value must be integrated into economic production and decision-making. In that sense, the OECD (2011a, p. 10) expects in its green growth strategy "mutually reinforcing aspects of economic and environmental policy" and emphasises the use of "cost-effective ways of attenuating environmental pressures to effect a transition towards new patterns of growth that will avoid crossing critical local, regional and global thresholds" (ibid.). Furthermore, economic growth would need to align with environmental policies ensuring "that natural assets continue to provide the resources and environmental services on which our well-being relies" (ibid., p. 9).[1] The OECD highlights that natural resources are often undervalued, which "imposes costs to the economy and human well-being" (OECD, 2011a, p. 17), or natural capital is valued but "only as a harvestable commodity and not for the vital ecosystem services provided" (OECD, 2020b, p. 7). Innovation is emphasised as it "can generate new sources of growth that better reflect the full value of natural capital to society and reduce the cost of addressing environmental risks" (OECD, 2011a, p. 12). However, key issues would include valuing not only natural resources, but all other "non-market benefits and costs, such as those related to health and life

[1] However, the OECD only indicates the empirical success of relative decoupling mainly in the field of material resource throughput (OECD, 2015a, p. 33; 2015c, p. 74; 2020d, p. 7).

expectancy" (OECD, 2011a, p. 19) as well, which again links to social issues. Furthermore, the OECD highlights "increasing efficiency" in relation to energy and resources. Generally, there is a focus on all kinds of technological improvement to ensure an increase in energy efficiency (OECD, 2017a, p. 52), with expected benefits also for social policy objectives in terms of employment gains (see above). In the field of industrial policy, the OECD promotes the idea of "industrial symbiosis", meaning that waste from one process should be reused for another (OECD, 2019a, p. 83). As far as policy instruments are concerned, the OECD highlights "market, regulatory, education and information-based instruments, as well as public financial support and co-operation across value chains" (OECD, 2020d, p. 12) that are supposed to drive efficiency. Closely connected with the latter, it promotes the idea of "increasing productivity", suggesting that "[i]ncentives for greater efficiency in the use of resources and natural assets" (OECD, 2011a, p. 9) be utilised as this would lead to an increase in productivity and a reduction of resource throughput (OECD, 2015c, p. 3). In that sense, the OECD suggests that environmental policy would not necessarily lead to a reduction of productivity: "in fact, tightening environmental policy stringency is associated with a subsequent short-run boost in productivity growth" (OECD, 2015a, p. 48).

As another related theme, the OECD promotes the concept of a "knowledge economy" or "weightless economy", one that is based on knowledge and services and does not rely on large amounts of resource throughput (OECD, 2017a, p. 153). In order to achieve this, the OECD supports research and development in green technologies and knowledge. Research and development are indeed supposed to be the "key driver[s] of innovation" (OECD, 2017a, p. 187) within a knowledge economy. Innovation would be particularly crucial in shifting farther back the frontier at which the use of natural capital impedes growth patterns: "By pushing the frontier outward, innovation can help to decouple growth from natural capital depletion" (OECD, 2011a, p. 10). In order to achieve this, the OECD prescribes various government practices that can support green innovation. Governments could, in particular, invest in research and provide finances for specific technological development, or they could enact "demand-side instruments such as standards, regulations and public procurement" (OECD, 2011a, p. 35). Governments could likewise promote the transfer and diffusion of low-carbon technologies, which would mainly be facilitated by "reducing barriers to trade and foreign direct investment, effective protection and enforcement of intellectual property rights, and efforts aimed at the least developed countries" (OECD, 2011a, p. 35).

Moreover, the OECD suggests that the implementation and development of low-carbon technologies within a green growth strategy could lead to "first-mover advantages" (OECD, 2017a, p. 152), accelerating economic growth for those who successfully provide the necessary innovations for a low-carbon economy, which can then be exported to other parts of the world.

In sum, the OECD problematises the incumbent economic growth regime and how it has affected the ecology and also fallen short in certain social policy prospects. The OECD explicitly highlights how green growth should be integrated with social policy considerations (for more on this, see the OECD's green Keynesianism understanding in Sect. 4.3.3) and, more generally, sets forth crucial themes of green growth as also discussed in the typology of Chap. 3. The next two subsections discuss green economy and green Keynesianism, both crucial perspectives that follow a green growth agenda. However, as already hinted in this section, the green Keynesianism perspective is the more dominant one in the OECD's discourse (see also the discussion in Chap. 6). The simultaneous existence of the conflicting perspectives of green economy and green Keynesianism is further explicated in Chap. 7 on the basis of Polanyi's concept of the double movement.

4.3.2 Market-Based Governance: Green Economy

While the previous subsection dealt with some more general themes of green growth that the OECD advances, this chapter reconstructs the elements of the OECD's policy discourse that have to do explicitly with the green economy perspective as set out in the eco-social policy typology (Chap. 3). Green economy is a green-growth-driven perspective that aims to increase the scope of private actors and favours voluntary adoption and self-regulatory practices by private actors as opposed to a regulatory top-down governance approach. In terms of eco-social potential, green economy is the most modest perspective in the typology that simply focuses on greening economic growth to enable traditional growth-dependent welfare state operations to continue. In Chap. 7 I argue that green economy is to a certain degree congruent with the first movement in Polanyi's double-movement concept.

The OECD places particular emphasis on propagating economic instruments such as carbon pricing or emission trading schemes to incorporate ecological externalities in economic decision-making, basically by giving the ecology and the climate an economic value. It has its own series on *Carbon Pricing and Energy Taxation* (OECD, 2021d) in which the latest developments in carbon pricing in several countries are discussed. It views these instruments favourably because they would incentivise innovation and present "a technology-neutral case for low-carbon investment and consumption, to promote green growth" (OECD, 2021b, p. 26). In that sense, the key rationale of carbon pricing is to increase "the benefit of carbon-neutral technologies making them worthwhile even in the absence of explicit green conditionality" (OECD, 2020f, p. 11). Moreover, with accompanying stimulus packages, carbon pricing can ensure that investments are directed towards decarbonisation because fossil-fuel-intensive investments would be too expensive (ibid.). Generally, there is a perception that economic tools of this kind are the most cost-effective means to guide actors towards low-carbon practices (OECD, 2011b, p. 25; 2017a, p. 198). The OECD argues too that these measures might "have knock-on effects on behaviours and activities not directly subject to the price" (OECD, 2017a, p. 189). But it criticises the fact that too many environmentally harmful activities, such as international aviation and maritime transport, domestic aviation and the use of coal, are not subject to carbon pricing mechanisms (OECD, 2019c, p. 10f). Furthermore, the OECD values economic instruments for their flexibility in achieving a reduction in emissions or ecological degradation. They also do not harm competition and innovation and are therefore considered to be market-friendly: "Taxes encourage behavioural change but do not force change when it would be excessively costly" (OECD, 2015d, p. 45). The OECD perceives economic instruments as universally applicable: "While national circumstances will differ, putting a price on pollution or on the over-exploitation of scarce natural resources [...] should be a central element of the policy mix" (OECD, 2011a, p. 12). In terms of social policy, economic instruments are highlighted as tools that generate revenues which could "help finance education, health care, infrastructure development or poverty alleviation" (ibid., p. 35). Comparing economic with regulatory instruments, the OECD states that the latter would "fail to provide an intrinsic mechanism for ensuring that environmental target be attained at the least economic cost" (ibid., p. 46). However, elsewhere, the OECD considers economic instruments as important supplements to regulatory approaches (OECD, 2020e, p. 14).

Another element of the OECD's green economy perspective is to favour free trade as crucial for transitioning to a low-carbon economy. The organisation promotes a reduction of "[b]arriers to trade and investment [that] can place a serious break on the development and diffusion of green technologies globally" (OECD, 2011a, p. 12). That would also be crucial in facilitating foreign direct investment and licensing technology (ibid.). In addition, the OECD encourages measures that provide information and disclosure to increase the transparency of sustainability practices. In that regard, the OECD highlights practices by "institutional investors to clarify and disclose climate change risks in regard to investment portfolios" (OECD, 2017a, p. 299). Connected to disclosure practices, the OECD supports various measures that focus on information provision, such as labelling schemes, or "voluntary and mandatory disclosure and reporting, mandatory energy audits and consumer protection policies" (ibid., p. 218). The OECD's promotion of information disclosures relates also to having knowledge of the economic value of natural resources or the ecological degradation that might result from certain practices. The organisation emphasises that such information might "support the functioning of the financial system and enhance decision-making about policy interventions" (ibid., p. 295). Closely connected to the general idea of information disclosure is "green budgeting". This has been particularly promoted by the documents *Green Budgeting in OECD Countries* (OECD, 2021b) and *Green Budget Tagging* (OECD, 2021c), both of which were compiled by the Public Management and Budget division in the Directorate for Public Governance. Green budgeting means that budgets are specifically labelled with regard to their ecological impact. This is supposed to "provide policy makers with a clearer understanding of the environmental and climate impacts of budgeting choices" (OECD, 2021b, p. 7). In that process, environmental impact assessments and cost-benefit analysis would be crucial (ibid., p. 8). The OECD appreciates green budget tagging as a "tool in raising awareness of how budget decisions are linked to national objectives" (OECD, 2021c, p. 3). It would increase transparency, give governmental budget decisions greater visibility as well as spreading knowledge about whether investments are sustainable or not (ibid.). However, the OECD also acknowledges that green budget tagging hardly changes policies. Instead, it facilitates informed budgeting choices in a way that might shift decision-making towards low-carbon and ecological protection objectives (ibid., p. 12).

As another crucial element of green economy, the OECD promotes a green finance system and prescribes "blended finance". The idea is to move towards a more climate-resilient finance system in which "possible misalignments within financial regulations and practices" (OECD, 2017a, p. 37) would need to be addressed. Furthermore, there would need to be more "diversified risk mitigation tools, improved environmental risk analysis, and diversified financial instruments and models" (ibid.) that aim to mobilise private capital. Along these lines, the OECD promotes "blended finance" which involves the mobilisation of private capital with the help of public investments and guarantees. Here, certain measures to decrease (private) risk are crucial, such as "guarantees, credit enhancements, currency hedging" (ibid., p. 34), insurance arrangements and "instruments such as green bonds and securitised loans [that] help to secure a reliable long-term funding basis for infrastructure projects" (ibid., p. 34f.). Overall, it would be crucial to "[i]ncrease potential for public finance to catalyse private investment" (OECD, 2020b, p. 8). For the OECD it will hardly be enough to mobilise public finance alone to meet climate mitigation objectives: "Governments need to target their scarce resources at mobilising large-scale private sector investment and at activities unlikely to attract sufficient private funding" (OECD et al., 2015, p. 51).

Connected to greening the finance system is the idea of a green sustainable taxonomy that the OECD promotes as well. Certain investments and practices would be labelled as either green or unsustainable in order to steer investments towards the green ones: "More precise and consistent definitions of which investments are 'green' and 'sustainable' could facilitate investment by giving confidence and assurance to investors" (OECD, 2020i, p. 11). Accordingly, a green taxonomy would serve to prevent "market fragmentation" and aim instead to provide "market clarity" (OECD, 2020i) while "potential users include corporates, financial institutions and retail investors" (ibid., p. 37). However, a taxonomy is not necessarily a mandatory guide for investment decisions, which resonates particularly well with the rather anti-regulatory approach of the overall green economy perspective:

> The assumption is that issuers of financial products will use the taxonomy in response to and anticipation of demand from investors, who will be motivated to make their portfolios, or a portion of their portfolios, more

consistent with the taxonomy, in view of reputational, risk- or opportunity-related considerations. (ibid., p. 39)

Taxonomies are also supposed to take account of broader, multi-dimensional well-being conceptions, which can include environmental as well as social safeguards (ibid., p. 45). Here, the OECD refers, for example, to its Guidelines for Multinational Enterprises, the UN Guiding Principles on Business and Human Rights, the International Labour Organisation's (ILO) Fundamental Rights and Principles at Work, the eight core conventions of the ILO and the International Bill of Human Rights.

In sum, the OECD aligns with crucial elements of the green economy perspective, as discussed in Chap. 3. This largely involves appreciating the capacity of private actors and abstaining from excessively regulating their behaviour. As discussed in Chap. 7, this reflects to some degree Polanyi's first movement. However, the OECD also promotes key elements of the countermovement, which is largely represented by green Keynesianism. The next subsection reconstructs the OECD's understanding of green Keynesianism.

4.3.3 The OECD as a Driver of Green Keynesianism: A Publicly Driven Transition

Here I reconstruct the elements of the OECD's policy discourse that relate to the green Keynesianism perspective as set out in Chap. 3's eco-social policy typology. Green Keynesianism, like green economy, is a green-growth-driven eco-social policy perspective, but unlike the latter, it aims to increase the scope of government intervention in order to regulate and control market developments and has more awareness of the ecological consequences of certain welfare state operations as well as the potential negative social effects of certain environmental policy constellations. The simultaneous existence of the green economy and green Keynesianism perspectives requires explanation (since they are antagonistic to each other). This is further explored in Chap. 7.

The OECD's green Keynesianism perspective entails various crucial elements explored in the typology that displays green Keynesianism as an eco-social policy perspective. Generally, the OECD promotes "green inclusive growth" (OECD, 2021a) with a focus on social protection while explicitly showing approval of government intervention. There is increased awareness of the relationship between environmental and social policy, recognising that they can be both regressive and synergistic with each other. This is connected to a comprehensive application of public investment measures that are earmarked for climate change and ecological protection objectives, including targeted support for green industries and technologies. Moreover, the OECD spreads awareness of how high-carbon lock-in effects can be prevented, for example in relation to infrastructure investments, and promotes the idea of "changing subsidies". Another crucial element is the OECD's circular economy perspective.

Several documents refer to elements of green Keynesianism. Important elements of it form part of the OECD's green growth strategy *Towards Green Growth* (OECD, 2011a) and the document *Investing in Climate. Investing in Growth* (OECD, 2017a). However, this section draws on several other documents as well, including NAEC's report on *Beyond Growth. New Approaches to Economic Challenges* (OECD, 2020a) and *The Inequalities–Environment Nexus. Towards a People-Centred Green Transition* (OECD, 2021a) which both involve key themes of green Keynesianism.

Green Keynesianism involves increased awareness of potential trade-offs and synergies between environmental and social policy (Sect. 3.2.2). This is also evident in the OECD's policy discourse. Among crucial heuristics put forward by the OECD to provide a framework for turning environmental and social policy trade-offs into synergies are the well-being lens, which is closely connected to the Centre for Well-being, Inclusion, Sustainability and Equal Opportunity (WISE), and systems thinking, which is connected to the NAEC initiative. In an edited volume entitled *Systems Thinking for Policy Making: The Potential of Systems Analysis for Addressing Global Policy Challenges in the 21st Century*, NAEC and the International Institute for Applied Systems Analysis present systems thinking:

> Systems analysis examines how complex problems interact, how threats and risks can multiply, and where feedbacks might exist, making it possible to anticipate surprises or tipping points. It provides a pathway to effective translation of research into impacts, by developing instruments to bridge sectors and actors, as well as temporal, social, and spatial scales from global to regional, national, and subnational, thus facilitating the task of policy- and decision-makers to address global and national challenges. (Hynes et al., 2020, p. 77)

In general, the OECD states that policies need to be integrated and therefore calls for an alignment of various policy fields with a climate change agenda. The report *Aligning Policies for a Low-Carbon Economy*, based on a collaboration between the OECD, the International Energy Agency, the International Transport Forum and the Nuclear Energy Agency (OECD et al., 2015), is an example of a perspective that focuses on integration and alignment of policies: "Better alignment will often facilitate the achievement of other policy objectives and can make climate policy more acceptable to various stakeholders [...]. Governments should identify and capitalise on these win-win outcomes" (ibid., p. 26). In this context, the OECD criticises the fact that many decision-making procedures and regulatory frameworks are traditionally not integrated or connected with the climate change policy framework but instead operate in policy silos: "Identifying and addressing these misalignments systematically in each country will enhance the responsiveness of economic and social systems to the climate change agenda" (ibid., p. 22). However, unsurprisingly given the collaborators who produced it, the report focuses not necessarily on social policy and climate policy alignment, but instead on the alignment of climate policy and issues of finance, taxation, trade policies, innovation and adaptation specifically in regard to the sectors of electricity, urban mobility and land-use. The alignment of social and environmental policy is best explored in *The Inequalities–Environment Nexus. Towards a People-Centred Green Transition* (OECD, 2021a). There, the OECD states that "incorporating measures addressing environmental sustainability, human health considerations and socioeconomic equity together in policy packages is more important than ever" (ibid., p. 36). Elsewhere, as part of the OECD's green growth strategy, the OECD emphasises that "[t]he potential social implications of green growth reform should be addressed ex ante, as a fundamental part of policy design, rather than as an afterthought to policy implementation" (OECD, 2015a, p. 64). Long-term strategies would need to make sure that there is a certain integration of different ministerial policy portfolios in which issues such as equity and environmental protection are mainstreamed in various different and traditionally separated policy decision-making processes (OECD, 2021a, p. 4). Moreover, the OECD calls for "inclusive green growth" to ensure that social protection and investments can be in place that "ease structural adjustment of local economies" (ibid.) within the transition to low-carbon economies.

Further exploring the relationship between environmental and social policy, the OECD emphasises the potential socially regressive effects of environmental and climate mitigation policies in various instances. It states that economic instruments (such as environmentally related taxes) that incorporate the ecological costs of certain products often have a regressive impact, meaning that low-income and socially vulnerable groups bear a disproportionate financial burden. Fuel-related taxes, for example, would be regressive, especially towards the elderly, low-income groups and households with children, as they spend a relatively large share of their income on heating and energy (OECD, 2006, p. 139; OECD et al., 2015, p. 39). The OECD (2015b, p. 45) also suggests that the removal of (fossil-fuel) subsidies usually has a regressive effect as well and emphasises, moreover, the potential regressive effect of comprehensive climate mitigation housing policies, which might make housing less affordable for low-income groups (OECD, 2019a, p. 112). In order to mitigate such burdens, it proposes changing income taxation and reducing taxation on wage labour, or other benefits that are connected to income as those usually support low-income groups in particular (OECD, 2011a, p. 102; OECD et al., 2015, p. 92). Other measures (that partly overlap with each other) suggested by the OECD include “transfer payments, revenue recycling and tax swapping, sometimes with revenues earmarked for social spending” (OECD, 2021a, p. 39) or “increases in basic allowance, introduction of non-wastable or wastable tax credits” (OECD, 2006, p. 22). Additionally, it proposes lump-sum transfers, whereby every household would receive the same amount of money. This approach would also disproportionately benefit lower-income households (OECD et al., 2015, p. 40). Accordingly, the OECD is aware of the potentially regressive social effects of environmental policy and emphasises the need to mitigate these effects through various compensation measures targeted at households and individuals, which is crucial for green Keynesianism considerations (Sect. 3.2.2).

In terms of industry compensation, the OECD proposes “bridge measures” to support older workers in declining industries as they transition into retirement, typically through income support. However, it also emphasises the need for incentives to discourage early retirement, as this could place a burden on fiscal sustainability (OECD, 2021a, p. 44). It also discusses the free allocation of emission allowances and other exemptions for certain industries but states that they need to be carefully designed so as not to “potentially create [...] barriers to entry of new and potentially more efficient firms" (OECD, 2011a, p. 109). And in order to capture compensation for affected workers in carbon-intensive industries, it deploys the concept of a just transition:

> Workers need a just transition to be ready for the green future. The green transition may create new jobs requiring new skills; however, some jobs, especially related to fossil fuels extraction and carbon-intensive heavy industries, could come under pressure. Upgrading workers' skills is vital to strengthen their resilience to risks and shocks, particularly in localities where economy is not well diversified and access to education is limited. (OECD, 2021a, p. 37)

Within a just transition to a low-carbon economy, the OECD expects that the development of green industries will create jobs, and there is no necessity to fear that there will be too few jobs if high-carbon industries decline. Instead, it anticipates that, in absolute terms, there will be more jobs due to the transition (OECD, 2019a, p. 48). It does, however, emphasise that there might be skill and knowledge gaps preventing workers from being able to benefit from these new jobs. Accordingly, its just transition approach focuses on labour market and education policies. "Minimising skill bottlenecks and facilitating the acquisition of new skills required of workers on both new jobs and existing jobs" (OECD, 2011a, p. 85); "Training and retraining programmes will be a key component of labour market policies" (ibid., p. 13). Moreover, in order to embark on a just transition, there would be a need to build partnerships between industry and governments (OECD, 2020c, p. 11). Within the context of a just transition, the OECD also discusses the potential for gendered influence as the workforce in traditional high-carbon industries is mostly male. Moreover, since there is low participation by women in the education of science and technology, there might be social barriers to women being part of the green transition on a technological level (OECD, 2021a, p. 4). Accordingly, the OECD has adopted a nuanced green Keynesian understanding that is closely connected to the notion of a "just transition of ecological modernisation" (Kreinin, 2020) as discussed in Sect. 3.2.2.

As also discussed in the same subsection, another crucial part of green Keynesianism is to promote an "entrepreneurial state", which is supposed to invest capital in the innovation and development of green and low-carbon industries and technologies (Goldstein & Tyfield, 2017). This idea is explicitly promoted by the OECD:

> states must become more entrepreneurial, seeking to shape markets and steer the process of economic change, not simply correct market failures. An empowering and entrepreneurial state of this kind would allow the development of a new kind of social contract—a new relationship between the state, business, civil society and citizens. (OECD, 2020a, p. 27f.)

In relation to recovery policies, the OECD also explicitly promotes the integration of decarbonisation objectives into such investments: "Designing recovery packages with decarbonisation objectives in mind will help ensure that the recovery is strong and builds a more sustainable growth trajectory than the past" (OECD, 2020f, p. 2). In fact, the OECD promotes fiscal spending as this might lead to "strong fiscal multipliers and job creation" (ibid.). Generally, the OECD states that the success of recovery should not be based on the extent of economic growth acceleration but on general increases in well-being, such as increases in income, job quality and housing quality instead (OECD, 2020b, p. 5).

Closely connected to an entrepreneurial state is the key green Keynesian notion of increasing government intervention within an economic framework to actively steer the market towards social and ecological objectives (Sect. 3.2.2). This approach is also supported by the OECD, which promotes various policy ideas that involve a relatively ambitious level of government intervention.

> But if public policy is to aim at different ethical outcomes, the state will have to play a larger role in guiding, or steering, the overall patterns of economic activity to achieve them. Through public service and welfare provision it can also support a fairer and more productive form of economic development. 'Correcting market failures' will not be sufficient; markets can also be 'shaped' in pursuit of publicly-determined goals. (OECD, 2020a, p. 23)

Moreover, the OECD states that "[g]overnment intervention is required to address well-known barriers and drive green innovation" (OECD, 2015a, p. 39). This is particularly the case where private actors do not have incentives to invest in green technology and practices. Government intervention could tackle potential barriers in existing technology markets that, due to path-dependency and lock-in effects, favour high-carbon practices and technology:

> The challenge is to gear innovation systems so that they both accelerate innovation generally and directly promote green technologies and processes,

through a 'system innovation' approach that addresses specific market failures, but also responds to demand-side issues such as consumer and household acceptance. (OECD, 2015a, p. 39)

The OECD also follows green Keynesianism in appreciating public investments while rejecting certain monetarist and austerity arguments (on green Keynesianism's rejection of monetarism and austerity, see Sect. 3.2.2). NAEC within the OECD notes that monetary policy can be insufficient in terms of recovery policies and guiding a transition towards a low-carbon economy: "a more active fiscal policy is both necessary and desirable in present conditions when interest rates are very low and monetary policy has largely run out of options" (OECD, 2020a, p. 21). Elsewhere, the OECD states that "little scope [is] left for monetary policy to provide such a stimulus in most countries, so a proactive fiscal policy response is required" (OECD, 2017a, p. 130). During economic shocks a fiscal policy framework would be crucial "that creates sufficient fiscal space during upturns to allow for a stimulating fiscal policy response during downturns and rapidly scaling up income support and active labour market programmes as needed" (OECD, 2018a, p. 15). To a certain degree, there is the tendency within the OECD's policy discourse to reject austerity measures and favour public borrowing: "it has been shown how high levels of public borrowing and debt can be sustained so long as the growth rate of the economy (which can itself be stimulated by public investment) exceeds the rate of interest paid. Public investment can 'crowd in', rather than 'crowd out', private finance" (OECD, 2020a, p. 21f.). Generally, the OECD encourages streamlining climate risk considerations into public investments. In so doing, governments should enact a more climate-resilient policy transition, particularly within procurement and lending practices (OECD, 2017a, p. 206). This speaks, for example, to infrastructure decisions, official development assistance and support for specific firms and research (OECD, 2011a, p. 57; 2017a, p. 220; OECD et al., 2015, p. 17). In relation to industry and technology, the OECD frames these as "mission-oriented" (OECD, 2020a, p. 26) industrial policy and "mission-oriented research programmes" (OECD, 2011a, p. 57). On that basis, the OECD prescribes making government support for industries conditional on ecological improvements and improved social and ecological resilience, which includes, for example, support for workers (OECD, 2020b, p. 8). This is also supposed to ensure "the long-term viability and competitiveness of firms in a low-carbon economy" (OECD, 2020g, p. 12).

Furthermore, the OECD (2020a, p. 26) states "that economic development has a direction as well as a rate". The direction should not be steered and controlled towards higher industrial performance per se but instead towards sustainability objectives. As policy measures to achieve this, the OECD mainly refers to public investments and procurement policies: "public demand can be geared towards supporting green industrial regeneration, green infrastructure development and supporting local job creation" (OECD, 2021a, p. 47). Generally, the organisation conceives a need for targeted public support in regard to low-carbon technologies: "The need for low-carbon innovation and the potential for spill-overs justify targeted support in R&D [research and development] and in technology-transfer towards application, including support through the corporate tax system" (OECD, 2020f, p. 13f.). Temporary support would be crucial to create a marketplace for such technology. However, "[i]nnovation support instruments must be carefully designed to foster the emergence and uptake of efficient technologies while minimising the risk of technology lock-in, lack of competition or crowding out of private investment" (OECD, 2011a, p. 12).

While pursuing public investments to support the overarching objective of decarbonisation, the OECD emphasises the importance of preventing negative lock-in effects on infrastructure. Governments would need to stop investing in carbon-intensive infrastructure because they would "risk locking in even higher levels of GHG emissions for decades to come, and they will enhance the risk of stranded assets" (OECD, 2017a, p. 106). Governments would also need to be cautious about the vulnerability of physical infrastructure with respect to climate change, especially if the infrastructure is supposed to be in operation for a long time (OECD, 2017a, p. 106). However, as the OECD admits, many important processes and technologies are not yet available that would ensure the prevention of lock-in effects for decades to come, especially within heavy industry. Therefore, in order to guard against future lock-in effects, "[d]emonstration and deployment to establish the commercial viability of such new technologies" (OECD, 2019a, p. 75) will be important. Potential regret for locking-in is particularly mentioned with regard to mobility infrastructure that supports individual car ownership as the main transportation mode (OECD, 2020b, p. 11; 2020c, p. 8f.).

As another crucial element of its green Keynesian understanding, the OECD promotes the idea of "changing subsidies". This acknowledges that fossil-fuel subsidies have a negative impact on well-being due to the air pollution and climate change impacts often related with them. Accordingly, governments should remove "perverse subsidies which encourage pollution or over-extraction of resources and place a drain on the public purse" (OECD, 2011a, p. 35). It states that "[f]ossil-fuel subsidies continue to serve as a major impediment to green growth, shoring up the role of incumbent polluting technologies, holding back investment in cleaner emerging technologies, and acting as a negative price on carbon" (OECD, 2015a, p. 15). Financial resources that would become available due to the withdrawal of subsidies could then "be used to alleviate energy affordability pressures on poorer households" (OECD, 2017a, p. 249), again emphasising the need for social compensation measures to combat the potentially regressive effects of fossil fuel subsidy removal (see also above; OECD, 2017a, p. 251). The newly available resources could also be utilised to support green technologies (OECD, 2011a, p. 42).

As far as the circular economy is concerned, which is also considered a crucial part of green Keynesianism (Sect. 3.2.2), the OECD has an understanding that consists mainly of promoting a more resource- and energy-efficient use of materials, the use of more sustainable and durable products in production processes and the incentivisation of the "three Rs": recycling, reusing and repairing (OECD, 2020d, p. 19). Overall, the OECD prescribes several policy instruments for advancing a circular economy "including economic instruments, green public purchasing, public information and awareness-raising measures, and private sector initiatives, such as voluntary agreements" (OECD, 2019b, p. 28). As far as regulation is concerned, the OECD is particularly in favour of "recycling targets, product standards, recycled content requirements, lifetime warranties, bans and restrictions and deposit-refund systems" (OECD, 2020d, p. 13). Circular economy is discussed particularly in relation to industry: "Heavy industry can only promote well-being if the broader economy shifts to a net-zero, circular and resource-efficient model" (OECD, 2019a, p. 78). Moreover, the OECD also considers the advantages of circular economy aspects in regard to value chains (OECD, 2020b, p. 12), declaring in short that "governments will need to shift away from the linear economy—where raw materials are extracted, processed, consumed and disposed—and mainstream decarbonisation, circularity and resource efficiency across the entire economy" (OECD, 2019a, p. 75).

In sum, the OECD promotes green Keynesian themes relatively prominently and explicitly links them to the synergistic achievement of ecological and social objectives while also being aware of certain regressive relationships between the policy fields. This contrasts with the green economy perspective that the OECD also promotes. However, green Keynesian themes predominate in the OECD's discourse compared to green economy ones. This paradoxical finding is further made sense of in Chap. 7 based on Polanyi's concept of the double movement (see Sect. 2.3).

4.3.4 *The OECD's Understanding of Recomposing Consumption and Production*

This subsection reconstructs the OECD's understanding of recomposing consumption and production. Based on the eco-social policy typology (Chap. 3), it summarises and structures how the OECD approaches the particular eco-social policy perspective of recomposing consumption and production. Generally, the OECD acknowledges that contemporary consumption and production patterns are unsustainable. In order to change these patterns, the OECD has a particular focus on changing norms, for example, with the help of public procurement policies. It generally looks at consumer practices from an individualistic point of view, focusing on consumer sovereignty. This includes empowering consumers by highlighting certification and labelling, economic instruments as well as educational and informational measures. Furthermore, it promotes extended producer responsibility. In general, the OECD's understanding of recomposing consumption and production is ambiguous. Some elements clearly speak to the green economy perspective, such as consumer sovereignty. Others are close to green Keynesianism, such as public procurement. This largely mirrors the double movement dynamic in the OECD's (and World Bank's) eco-social policy discourse(s), which is further discussed in Chap. 7. It cannot be said that the OECD's understanding of recomposing consumption and production sets the stage for post-growth.

Several OECD documents that represent the core of its take on recomposing consumption and production require attention. Most of these are connected in one way or another with the green growth strategy, and several documents already discussed in the previous subsection reappear, but a number of earlier documents published before the green growth strategy are crucial to the OECD's understanding of recomposing consumption and production as well. For example, *The Political Economy of Environmentally Related Taxes* (OECD, 2006) addresses various themes related to consumption. This work was a collaboration between the Centre for Tax Policy and Administration and the OECD Environment

Directorate. Other key documents include *Creating Incentives for Greener Products* (OECD, 2015d), produced with external collaboration but substantially under the leadership of the Environment Directorate; *Tackling Environmental Problems with the Help of Behavioural Insights* (OECD, 2017b), developed under the leadership of the Environment and Economy Integration Division within the Environment Directorate; and the *Consumer Policy Toolkit* (OECD, 2010), which was based on the work of the Committee on Consumer Policy. The strong connection to the green growth strategy suggests that its understanding of recomposing consumption and production is closely aligned with green growth (for a detailed discussion of the relationship between green growth and recomposing consumption and production, see Sect. 3.5). However, one notable document outside the green growth stream is *Opportunities for All* (OECD, 2018b), which includes general concepts related to recomposing consumption and production. This document was developed and coordinated as part of the inclusive growth stream, involving the contributions of several different directorates.

The OECD acknowledges that "there is growing consensus that the current consumption path is not environmentally sustainable" (OECD, 2015c, p. 46). In terms of policy instruments in relation to recomposing consumption and production, the OECD has a very broad understanding: "any policy instrument used to achieve environmental targets should cause changes in consumption and/or production patterns" (OECD, 2006, p. 16). Following this, the OECD highlights the question of "who should change their behaviour, by how much and within which timeframe" (OECD, 2006, p. 16). Later it emphasises that the 10 per cent of richest people in the world are responsible for around 50 per cent of individual lifestyle emissions. This, at least implicitly, brings the issue of redistributing consumption and natural resources to the discussion (OECD, 2018b, p. 141). The OECD also states that the consequences of climate change and ecological degradation affect most those who have least contributed to them (ibid.). The OECD explicitly calls for resource use by high-income groups to be reduced. Policy measures should target those that consume "the greatest quantities of material resources, who typically come from the upper end of the income distribution" (ibid., p. 161). Furthermore, the OECD acknowledges that the territorial-based way of accounting GHG emissions "can mask climate impacts of production systems and consumption choices that are inefficient from a GHG emissions perspective" (OECD, 2020e, p. 24). Moreover, it can obscure the understanding of emissions from certain production systems. When viewed within a territorial-based accounting system, these systems may seem more sustainable compared to how they would appear under a CBA system (ibid.). Closely connected to green growth, the OECD appreciates innovation that can create new production and consumption practices: "A strong capability to innovate is essential to establish the capacity for breakthroughs and new patterns of production and consumption" (OECD, 2011a, p. 12). It also prescribes certain prohibitions, in particular "those designed to protect social values" (OECD, 2010, p. 99) as they are helpful in achieving larger societal objectives. Generally, the OECD also recommends supply measures to support sustainable production and consumption practices. Here, the OECD mentions "collection services for recyclable materials, the provision of public transportation, or the characteristics of electricity supply" (OECD, 2011b, p. 17).

There is a certain focus on discussing the influence of culture and norms when it comes to changing consumption and production patterns. As the green growth strategy states: "consumers are to some extent locked into norms and habits of consumption and policy needs to reflect on ways of enabling consumer change" (OECD, 2011a, p. 36). Similar ideas are present in *Tackling Environmental Problems with the Help of Behavioural Insights* (OECD, 2017b). Accordingly, individuals tend to compare themselves with others, meaning that consumption decisions and practices are not always driven by rational calculations. Instead, these choices are often influenced by how others perceive them and whether they are considered socially appropriate (OECD, 2017b, p. 13). For example, the decision to buy a car is frequently shaped more by its value as a status symbol than by considerations of fuel efficiency (OECD, 2017b, p. 92). To change norms, the OECD suggests, for example, the use of information-based policy instruments (OECD, 2011b, p. 30). Moreover, it prescribes "nudging instruments" that focus on changing routine practices, largely by changing default options. This simply acknowledges that consumption decisions are often biased towards the status quo: "simplifying their daily effortful activities by changing default options to more environmentally sustainable ones can 'green' routine behaviour without entailing additional effort on the part of consumers" (OECD, 2017b, p. 133). This approach may be effective, as individuals often tend to choose the easiest option and avoid decision-making about what might be the best choice (OECD, 2010, p. 46). Another suggestion from the OECD for changing routine behaviour is to provide timely feedback on the impact of consumption decisions, such as real-time monitoring of household electricity use (OECD, 2017b, p. 13).

As a stronger demand-steering policy tool, the OECD suggests public procurement, which is in particular highlighted in relation to industrial renewal. Here, public demand could be steered "towards supporting green industrial regeneration, green infrastructure development and supporting local job creation" (OECD, 2021a, p. 47). Public procurement is highlighted in terms of creating new markets and enabling first-mover advantages by creating "lead markets" (OECD, 2017a, p. 32) for innovation. Measures that the OECD suggests are to price emissions within procurement criteria (ibid.), or to integrate resource efficiency criteria in procurement practices (OECD, 2020d, p. 16). Another way to guide public procurement practices would be the use of eco-labels or to use procurement to support the circulation of circular goods (circular procurement) (OECD, 2019b, p. 73).

However, the OECD places strong responsibilities on individual consumers and favours a policy-narrative of enabling consumers that focuses on consumer sovereignty: "When they are empowered, consumers can improve economic performance by helping to drive competition and business innovation" (OECD, 2010, p. 8). Hence, consumers are supposed to drive themes that are also archetypical of green growth: "innovation, productivity and growth" (OECD, 2010, p. 3). The OECD promotes programmes "that can help consumers make better decisions, by promoting market transparency and by providing information and guidance" (ibid., p. 28). "Competitive markets" are also promoted as they are considered to be "the best way to ensure an efficient level of information for consumers" (OECD, 2010, p. 33). Generally, the OECD prefers a policy instrument mix made up of taxes and regulations which would "be preferable to a blanket ban on the more damaging options, which can be unduly inflexible, and may deny consumers products which, for a variety of reasons, they may value strongly" (OECD, 2015d, p. 24). This rejects more interventionist policies, such as consumption corridors (Fuchs et al., 2021).

As another theme related to enabling consumers and consumer sovereignty, the OECD suggests the application of certification and labelling, particularly in relation to food, to help educate consumers and promote healthy and sustainable consumption choices (OECD, 2019a, p. 180). To influence businesses, the OECD emphasises labelling schemes that display energy consumption rates and the carbon footprint of specific production processes (OECD, 2017a, p. 218). Third-party certification is also encouraged, as it may be perceived as more independent, thereby fostering greater confidence in the social and environmental sustainability of a product. Moreover, the OECD suggests that governments can take the initiative by creating comparative information schemes that enable consumers to make well-informed decisions (OECD, 2011a, p. 50). Labelling schemes would be particularly effective when "there is an underlying demand for environmental quality" (OECD, 2011b, p. 26). These labels allow consumers to make choices in line with their environmental attitudes by providing information about the environmental impacts of consumption decisions and production processes (ibid.). However, the OECD stresses that trust in these labels must be established (ibid., p. 151). Additionally, labelling can help explain to consumers why higher taxes may apply to certain products, due to tax differentiation (OECD, 2015d, p. 24).

Furthermore, the OECD discusses the value and impact of economic instruments in relation to changing consumption and production patterns. They are supposed to "shift producer and consumer behaviour towards more environmentally beneficial activities and products" (OECD, 2015a, p. 34). The OECD explains that they function in such a way that a "higher tax will increase the price of 'dirty' goods to consumers, and reduce the profitability of producers of these goods, encouraging both to shift to cleaner alternatives" (OECD, 2015d, p. 10). While the OECD values economic instruments for influencing consumer and production practices due to their alignment with market-based governance, it also recognises that these measures alone may be insufficient to "guarantee the scale of the behavioural response that will be achieved" (ibid., p. 19). Moreover, economic instruments are supposed to be designed to work effectively in synergy with other policy measures, such as product standards (ibid., p. 24). The OECD highlights that a key advantage of economic instruments would be their cost-effectiveness, as well as the flexibility they provide in shaping consumer and producer responses: They "encourage behavioural change but do not force change when it would be excessively costly" (ibid., p. 45). Additionally, the OECD advocates the use of tax differentiation, where the tax structure depends on the environmental and social impact of the taxed good or service. In this way, the tax reflects the environmental and social costs of the product. The aim is to impose higher taxes on goods and services that are socially or environmentally harmful, while promoting the consumption of more sustainable alternatives. However, the OECD notes that the potential revenue generated from such a tax system is likely to be modest (ibid., p. 12).

In relation to certification and labelling, the OECD more generally prescribes these policy instruments to increase information provision and also as a supplement to economic ones. Increased information provision about the social and environmental aspects of a product might strengthen "markets for environmentally friendly products, particularly for goods and services where environmental attributes are less 'visible'" (OECD, 2011a, p. 50). Information provision would also highlight "the negative environmental consequences of specific activities, the long-term financial savings from using greener products and the availability of cleaner alternatives" (OECD, 2015a, p. 23). Overall, governments can either provide information about certain products themselves or oblige producers to provide it or encourage them to provide it on a voluntary basis (OECD, 2010, p. 83). The OECD also considers information provision as a general method protecting consumers against harmful or deceptive practices (OECD, 2010, p. 82). In order to receive relatively independent information, the OECD suggests increasing the status of intermediaries and third-party experts as they might be suitable for translating complex descriptions or instructions into easy-to-understand information (OECD, 2010, p. 84). Moreover, informational measures are considered important for raising consumer awareness about the consequences of their decisions and increasing political acceptance of policies that promote a transition towards more sustainable practices (OECD, 2011b, p. 16). In the field of food policy, education is regarded as crucial for informing people about the health and environmental benefits of their food choices (OECD, 2019a, p. 187). Additionally, education is particularly valued for its "important role in forming attitudes to environmental issues from early childhood on" (OECD, 2011a, p. 50), which can ultimately contribute to changing consumer behaviours. In fact, consumer education is considered to be the "foundation for other consumer policy tools" (OECD, 2010, p. 79).

As another theme, the OECD discusses the value of extended producer responsibility. Here, particularly, the document *Extended Producer Responsibility. Updated Guidance for Efficient Waste Management* (OECD, 2016), which was created by the Environment and Economy Integration Division, is crucial. In extended producer responsibility programmes, producers would be responsible for their products after they have been used by consumers (ibid., p. 21). Extended responsibility policies can entail a variety of instruments, including "take-back requirements, recycling targets, producer fees – that, collectively, make producers responsible for products at the end of their life" (OECD, 2019b, p. 88). In such programmes, producers are usually free to decide whether to simply provide financial resources to other actors that then take care of products after their use-cycle, or to organise the extended producer responsibility completely by themselves. Even though they create a certain burden on producers, the promotion of flexibility would still be key, leaving "producers, either individually or collectively, with more flexibility to determine how they achieve the objectives of the policy" (OECD, 2015d, p. 68). The OECD notes that, on a conceptual level, extended producer responsibility might actually come with a reduction in government interventions since it is left relatively open how producers comply with their responsibility. But governments would still be crucial owing to the general "requirement for monitoring and enforcement of public policy" (OECD, 2016, p. 65), which also suggests a link to green economy.

In sum, the OECD's conception of recomposing consumption and production reflects a certain dualism. While there is an appreciation for some public modes of consumption, there is also a strong emphasis on individualistic policy solutions that target the "empowered consumer" and promote market dynamics. This duality supports the argument presented in Sect. 7.1, which suggests that the OECD's (and the World Bank's) eco-social policies are characterised by a double movement dynamic: one movement focusing on expanding the role of individual private actors and market-based solutions in establishing green, self-regulating markets, while the countermovement emphasises government intervention, social protection and a greater focus on community-oriented, communal forms of ownership. It cannot, however, be said that the OECD's understanding of recomposing consumption and production serves as a transition to post-growth (see for this potential relationship Sect. 3.5).

4.3.5 *The OECD's Approach to Post-Growth: Sceptical Notions on GDP and Beyond*

In order to have a full picture of the OECD's eco-social policy perspective, it is necessary to reconstruct the OECD's post-growth understanding as well. Post-growth is the most transformative perspective in the eco-social policy typology (Chap. 3). It is an explicit counterperspective to green growth; it states that rejecting economic growth in the Global North would be inevitable in order to comply with planetary boundaries and enable development of the Global South that would not endanger ecological thresholds. This subsection illustrates that there are some elements of post-growth in the OECD's policy discourse but its post-growth understanding can only be reckoned to be marginally developed. The OECD picks up some notions that tend towards post-growth (especially critical remarks on GDP as an indicator of social progress) but then applies these arguments to promoting green growth, inclusive growth or "beyond growth" perspectives. Generally, the OECD displays overarching scepticism on applying GDP as a proxy for well-being and as *the* objective of policy-making. Moreover, it employs a multidimensional well-being framework and suggests going "beyond growth" but not to the extent of actually abandoning it. It refers to some policy ideas that are often discussed in post-growth circles, such as job guarantees or working time reduction policies as well as universal basic services, but this only takes place at the margin of its policy discourse.

Among the crucial documents for the OECD's post-growth perspective is, for example, *How's Life? Measuring Well-Being* (OECD, 2011c; more recently OECD, 2020h) created by the Well-Being Unit of the OECD Directorate of Statistics. The report was launched with the OECD's Better Life initiative. It introduces indicators that cover different dimensions that are all important for well-being, including housing conditions, health, education and work–life balance. Every two to three years, the OECD publishes updates as part of its Better Life initiative. Moreover, NAEC's document on *Beyond Growth. Towards a New Economic Approach* (OECD, 2020a) also entails post-growth ideas. Beyond that, some post-growth ideas are in fact part of the OECD's green growth documents.

The OECD has promoted criticism of GDP. In several publications, it emphasises that GDP is not a good proxy for well-being, noting that "it is now widely recognised that data on GDP provide only a partial perspective on the broad range of factors that matter to people's lives" (OECD, 2011c, p. 14). With regard to ecological protection, it highlights that growth in GDP "does not reflect environmental externalities associated with production and consumption", similarly that it would not reflect the "depreciation, depletion and degradation of [environmental] assets" (OECD, 2011a, p. 114). Moreover, there exist "regrettable" or bad practices that can increase GDP but undeniably reduce people's well-being. This is, for example, the case with increased congestion or increased spending on pollution abatement: "These 'regrettables' contribute to high economic activity but they obviously do not add to people's well-being" (OECD, 2011c, p. 17). The OECD creates comprehensive lists of issues that would not be captured by GDP. These include "people's health status, their happiness, their personal security and their social connections, all of which matter to people independently of their effect on people's consumption possibilities" (ibid.). Since these aspects are not conveyed or achieved through market interactions, they are not captured by GDP, and their evaluation relies on non-market measures (ibid.). Furthermore, GDP falls short in reflecting how income is distributed, but understanding income distribution is crucial for assessing societal well-being. Additionally, GDP does not account for services that lack a monetary value but are essential for supporting economic activities. Care work in households, for example, is one such service (OECD, 2019a, p. 25). Overall, the OECD summarises that

> [i]t is now well known that GDP is not a good measure of overall economic performance. It does not take any account of the distribution of income and wealth; it captures only flows of income not the stocks of capital that generate them; it undervalues unpriced and intangible services; it ignores unpaid work; it fails to measure environmental degradation; it is not a good proxy for wellbeing. (OECD, 2020a, p. 17)

For its part, the OECD promotes a well-being framework that argues that it goes "beyond growth" and "focuses on people's conception of well-being rather than the economic system or GDP" (OECD, 2014, p. 9). NAEC likewise affirms that "politicians and policymakers need to go 'beyond growth'. They need to ensure that, alongside rising GDP, economic policy is achieving wider objectives and measures of economic and social progress" (OECD, 2020a, p. 17). The well-being framework provides an integrative perspective covering various dimensions that are crucial for well-being: "income, jobs, health, knowledge and skills, safety and the quality of the environment, as well as the economic, natural, human and social capital stocks needed to sustain well-being over time" (OECD, 2019a, p. 11). This perspective is crucial to making "synergies and trade-offs systematically visible" (ibid., p. 3). Applying a "more comprehensive set of indicators can help monitor and set criteria to ensure progress on multiple policy priorities" (ibid., p. 4) and would be crucial to going "beyond growth". It would also be important to reject a silo approach in policy-making (OECD, 2019a). Elsewhere, though, the OECD frames going beyond growth as developing a "people-centred growth model"

(OECD, 2018b, p. 3), which means that economic growth is not actually rejected.

The OECD refers to some prominent post-growth policy ideas. It emphasises "the potential of reducing working hours to capture the gains of higher productivity in improved wellbeing rather than simply higher consumption" (OECD, 2020a, p. 27). It emphasises productivity gains due to automation and calls for "ensuring that the benefits of higher productivity do not accrue simply to the owners of capital" (ibid.). It also notes in that regard the potential of job guarantees that might level the effects of automation and briefly refers to universal basic services (ibid.).

In sum, there are some instances in which the OECD prescribes post-growth elements as discussed in the eco-social policy typology of Chap. 3, but uses these ideas in order to advocate for economic growth as a persistent policy objective, framing it as green growth, inclusive growth or going beyond growth. These contestation processes concerning the "economic growth paradigm" are not new, however. As Schmelzer (2016) suggests, the contestation of new issues and their integration into the economic growth paradigm indicate that the paradigm is not a stable set of discourses. Historically,

> it was continuously renegotiated and remade in an open and contingent process characterised by historical ruptures, competing theories, and counter-currents in which the growth paradigm proved remarkably flexible in adapting to changing circumstances, integrating newly emerging problems and perspectives without changing its basic tenets. (Schmelzer, 2016, p. 14)

The integration of certain post-growth concepts appears to be a continuation of this trend, as the OECD also continues to strongly advocate for economic growth (OECD, 2023; Schulze Waltrup, 2024). More radical post-growth policy proposals such as community currencies, non-monetary exchange systems, alternative credit institutions and redistributive policies such as income and capital caps, a citizen's income and new ownership patterns are absent from the OECD's policy discourse.

References

Barde, J.-P. (2007). Harnessing the political economy of environmental policy: David Pearce's contribution to OECD. *Environmental and Resource Economics, 37*(1), 33–42. https://doi.org/10.1007/s10640-007-9113-8

Bernstein, S. (2001). *The compromise of liberal environmentalism.* Columbia University Press.

Borowy, I. (2017a). (Re-)thinking environment and economy: The organisation for economic co-operation and development and sustainable development. In W. Kaiser & J.-H. Meyer (Eds.), *International organizations in environmental protection. Conservation and globalization in the twentieth century* (pp. 211–240). Berghahn Books.

Borowy, I. (2017b). Negotiating environment: The making of the OECD environment committee and the polluter pays principle, 1968–1972. In M. Leimgruber & M. Schmelzer (Eds.), *The OECD and the international political economy since 1948* (pp. 311–334). Palgrave Macmillan.

Busch, P.-O. (2009). The OECD environmental directorate: The art of persuasion and its limitations. In F. Biermann & B. Siebenhüner (Eds.), *Managers of global change: The influence of international environmental bureaucracies* (pp. 75–99). MIT Press.

Deacon, B. (2007). *Global social policy & governance.* Sage.

Deeming, C., & Smyth, P. E. (Eds.). (2018). *Reframing global social policy. Social investment for sustainable and inclusive growth.* Policy Press.

Fergusson, R. (2021). International organizations' involvement in youth unemployment as a global policy field, and the global financial crisis. In D. Niemann, K. Martens, & A. Kaasch (Eds.), *International organizations in global social governance* (pp. 31–56). Palgrave Macmillan.

Fuchs, D., Sahakian, M., Gumbert, T., Di Giulio, A., Maniates, M., Lorek, S., & Graf, A. (2021). *Consumption corridors. Living a good life within sustainable limits.* Routledge.

Goldstein, J., & Tyfield, D. (2017). Green Keynesianism: Bringing the entrepreneurial state back in(to question)? *Science as Culture, 27*(1), 74–97. https://doi.org/10.1080/09505431.2017.1346598

Hynes, W., Lees, M., & Müller, J. M. (Eds.). (2020). *Systemic thinking for policy making. The potential of systems analysis for addressing global policy challenges in the 21st century.* OECD Publishing.

Jenson, J. (2010). Diffusing ideas for after neoliberalism. *Global Social Policy, 10*(1), 59–84. https://doi.org/10.1177/1468018109354813

Kaasch, A. (2015). *Shaping Global Health Policy. Global social policy actors and ideas about health care systems.* Palgrave Macmillan.

Kreinin, H. (2020). *Typologies of 'just transitions': Towards social-ecological transformation*. Ecological Economic Papers, 35., WU Vienna University of Economics and Business, Vienna.

Lehtonen, M. (2007). Environmental policy integration through OECD peer reviews: Integrating the economy with the environment or the environment with the economy? *Environmental Politics, 16*(1), 15–35. https://doi.org/10.1080/09644010601073432

Long, B. L. (2000). *International environmental issues and the OECD 1950–2000. An historical perspective*. OECD.

Mahon, R. (2009). The OECD'S discourse on the reconciliation of work and family life. *Global Social Policy, 9*(2), 183–204. https://doi.org/10.1177/1468018109104625

Mahon, R. (2010). After neo-liberalism? The OECD, the World Bank and the child. *Global Social Policy, 10*(2), 172–192. https://doi.org/10.1177/1468018110366615

Mahon, R. (2015). The OECD, gender, and social policy—Parallel stories. In *Actors and Agency in Global Social Governance* (pp. 81–101). Oxford University Press.

Mahon, R. (2019). Broadening the social investment agenda: The OECD, the World Bank and inclusive growth. *Global Social Policy, 19*(1–2), 121–138. https://doi.org/10.1177/1468018119826404

Mahon, R., & McBride, S. (2009). Standardizing and disseminating knowledge: The role of the OECD in global governance. *European Political Science Review, 1*(1), 83–101.

McBride, S., & Williams, R. A. (2001). Globalization, the restructuring of labour markets and policy convergence. *Global Social Policy, 1*(3), 281–309. https://doi.org/10.1177/146801810100100302

Niemann, D., & Martens, K. (2018). Soft governance by hard fact? The OECD as a knowledge broker in education policy. *Global Social Policy, 18*(3), 267–283. https://doi.org/10.1177/1468018118794076

OECD. (1981). *The crisis of the welfare state*. OECD Publishing.

OECD. (1994). *New orientations for social policy*. OECD Publishing.

OECD. (2006). *The political economy of environmentally related taxes*. OECD Publishing.

OECD. (2008). Growing unequal. In *Income distribution and poverty in OECD countries*. OECD Publishing.

OECD. (2010). *Consumer policy toolkit*. OECD Publishing.

OECD. (2011a). *Towards green growth*. OECD Publishing.

OECD. (2011b). *Greening household behaviour: The role of public policy*. OECD Publishing.

OECD. (2011c). *How's life? Measuring well-being*. OECD Publishing.

OECD. (2011d). *Divided we stand. Why inequality keeps growing.* OECD Publishing.

OECD. (2014). *Green growth and sustainable development forum.* Conference Summary Report. Addressing the social implications of green growth. OECD Publishing.

OECD. (2015a). *Towards green growth? Tracking progress.* OECD Publishing.

OECD. (2015b). *Final NAEC synthesis. New approaches to economic challenges.* OECD Publishing.

OECD. (2015c). *Material resources, productivity and the environment.* OECD Publishing.

OECD. (2015d). *Creating incentives for greener products: A policy manual for eastern partnership countries.* OECD Publishing.

OECD. (2015e). *In it together. Why less inequality benefits all.* OECD Publishing.

OECD. (2016). *Extended producer responsibility: Updated guidance for efficient waste management.* OECD Publishing.

OECD. (2017a). *Investing in climate, investing in growth.* OECD Publishing.

OECD. (2017b). *Tackling environmental problems with the help of behavioural insights.* OECD Publishing.

OECD. (2018a). *Good jobs for all in a changing world of work: The OECD jobs strategy.* OECD Publishing.

OECD. (2018b). *Opportunities for all: A framework for policy action on inclusive growth.* OECD Publishing.

OECD. (2019a). *Accelerating climate action: Refocusing policies through a well-being lens.* OECD Publishing.

OECD. (2019b). *Waste management and the circular economy in selected OECD countries: Evidence from environmental performance reviews.* OECD Publishing.

OECD. (2019c). *Taxing energy use 2019: Using taxes for climate action.* OECD Publishing.

OECD. (2020a). *Beyond growth: New approaches to economic challenges.* OECD Publishing.

OECD. (2020b). *Building back better. A sustainable, resilient recovery after COVID-19.* OECD Publishing.

OECD. (2020c). *Making the green recovery work for jobs, income and growth.* OECD Publishing.

OECD. (2020d). *Improving resource efficiency and the circularity of economies for a greener world.* OECD Publishing.

OECD. (2020e). *Towards sustainable land use: Aligning biodiversity, climate and food policies.* OECD Publishing.

OECD. (2020f). *Green budgeting and tax policy tools to support a green recovery.* OECD Publishing.

OECD. (2020g). *COVID–19 and the low-carbon transition. Impacts and possible policy responses.* OECD Publishing.

OECD. (2020h). *How's life? 2020: Measuring well-being*. OECD Publishing.

OECD. (2020i). *Developing sustainable finance definitions and taxonomies*. OECD Publishing.

OECD. (2021a). *The inequality/environment nexus: Towards a people-centred green transition*. OECD Publishing.

OECD. (2021b). *Green budgeting in OECD countries*. OECD Publishing.

OECD. (2021c). *Green budget tagging: Introductory guidance & principles*. OECD Publishing.

OECD. (2021d). *Effective carbon rates 2021. Pricing carbon emissions through taxes and emissions trading*. OECD Publishing.

OECD. (2023). *OECD economic outlook, interim report March 2023: A fragile recovery*. OECD Publishing.

OECD, IEA, NEA and ITF. (2015). *Aligning policies for a low-carbon economy*. OECD Publishing.

Schmelzer, M. (2016). *The hegemony of growth: The OECD and the making of growth*. Cambridge University Press.

Schulze Waltrup, R. (2024). The OECD, global challenges and contestation of the economic growth paradigm. In A. Isakova, M. Neuwinger, R. Schulze Waltrup, & O. Uraiqat (Eds.), *Constructing global challenges in world politics* (pp. 214–227). Routledge.

Smyth, P., & Deeming, C. (2018). Conclusion: Towards a new global social policy framework? In C. Deeming & P. Smyth (Eds.), *Reframing global social policy. Social investment for sustainable and inclusive growth* (pp. 323–342). Policy Press.

Woodward, R. (2009). *The Organization for Economic Co-operation and Development (OECD)*. Routledge.

CHAPTER 5

The World Bank's Eco-Social Policy Discourse

This chapter provides a reconstruction of the World Bank's eco-social policy discourse. As in the previous chapter on the OECD, the next two sections provide an overview of the World Bank's involvement in social and environmental policy. This review is contextualised within the framework of Polanyi's double movement, which is a crucial policy dynamic that shapes the OECD's and World Bank's eco-social policy-making (see Sect. 7.1). However, this dynamic is present not only within eco-social policy but also in the development of social and environmental policy. The third section of this chapter provides an empirical reconstruction of the World Bank's eco-social policy discourse, for which the eco-social policy typology developed in Chap. 3 serves as the analytical framework. The chapter begins, however, with a brief introduction to the World Bank.

In terms of its organisational structure, the World Bank Group consists of five branches: the International Bank for Reconstruction and Development, which provides financing to low- and middle-income countries (commonly referred to as the World Bank); the International Development Association, which provides financing to low-income countries; the International Finance Corporation, which provides financing to the private sector; the Multilateral Investment Guarantee Agency, which encourages private businesses to invest in foreign countries; and the International Centre for Settlement of Investment Disputes, which helps resolve conflicts between private investors and foreign countries. In this

R. Schulze Waltrup, *Eco-Social Policy in the OECD and the World Bank*, https://doi.org/10.1007/978-3-031-95429-0_5

book, the primary focus is on the International Bank for Reconstruction and Development. Most of the documents analysed in this chapter originate from units and departments of the International Bank for Reconstruction and Development or reflect the approach of the entire World Bank Group.

While the World Bank also has financial capacities that enable it to finance, for example, infrastructure projects in client countries and to achieve policy-conditionalities based on the loans that it can provide (triggering policy reforms thanks to loans), this book largely focuses on the World Bank on the basis of its own conception of itself as a "knowledge bank" (Stone, 2003). Like the OECD, the World Bank influences policy discourses based on its expertise and data collection, aiming to institutionalise certain conceptions of policy best practices. This book follows this focus by examining the ideas and concepts promoted by the World Bank in terms of eco-social policy integration. Within it, various units work on selected topics, such as social development, climate change and economic policy. Major decisions are generally made by the member-states represented on the Board of Governors, where the US holds veto power over major decisions. Developed countries typically control large shares of the voting power, while developing countries, despite being client countries and being the World Bank's focus in terms of policy support, do not (Vetterlein & Schmidtke, 2024). As Chap. 6 will illustrate, some differences between the OECD's and the World Bank's eco-social policy can be explained by the fact that the latter mainly prescribes policies to low-income and developing countries (while the former focuses on high-income countries).

5.1 The World Bank in Global Social Policy

To contextualise the discussion of the World Bank's eco-social policy discourse, this section surveys social policy developments within the Bank. First, it reviews various studies that provide a general overview of certain eras of policy-making within the Bank. Second, it focuses more closely on studies of more specific social policy fields to understand how such large-scale developments unfold in more specific policy debates.

Over the decades, the World Bank has established various practices that can be subsumed under the term of social policy. First, it has made investments in key social sectors, including health, education and pensions. Second, it prioritises poverty eradication as a key organisational objective, as exemplified by its promotion of safety nets. Third, it applies different technical analyses, including social assessments, poverty and social impact analysis, and publishes papers that discuss how to best address poverty. Fourth, it has established social safeguards, trying to ensure that its infrastructure projects do not become harmful (Hall, 2007, p. 152). Generally, the term "social policy" has been used rather vaguely and infrequently within the World Bank. Various terms have been applied instead, resulting in a diverse and diffuse understanding of social policy: This "reflect[s] historical accumulation of a diverse social agenda within the Bank during its lifetime, reflecting critical junctures in policy making" (ibid.). This diversity is also reflected in the following accounts of the World Bank's social policy and can be made sense of based on Polanyi's concept of the double movement (see Sects. 2.3 and 7.1).

Bebbington et al. (2004; in reference to Kapur et al., 1997) identify four different policy eras. These eras represent different configurations of global social policy arrangements in the World Bank, and they come with "different problem definitions and underlying assumptions, and hence, different policy responses" (Vetterlein, 2007, p. 154). The first, in the aftermath of World War II, was largely about engineering and creating physical infrastructure. This came with investments in roads, electricity and ports. While these physical developments had at least implicit social considerations, it was not necessarily on the World Bank's agenda to shape policy discourse on social policy. The second era was then about putting poverty explicitly at the centre of development, an idea that was forcefully pushed by the Robert McNamara presidency (1968–1981). The focus was on investments in rural development to mitigate poverty so that countries would not feel tempted to turn to communist ideology. In the third era, starting from the 1980s, development was about policy reform as part of structural adjustment programmes. Structural adjustment is characterised by policies of deregulation, privatisation and (trade) liberalisation. This approach is accompanied by a focus on macroeconomic stability, aimed at reducing and preventing fiscal deficits, typically achieved through cuts in public spending and the use of monetary policies to control inflation. The economy is liberalised to encourage foreign direct investment, allowing for freer flows of capital and trade, while domestic markets are deregulated and privatised. In general, structural adjustment policies envision a limited role for government intervention, which is primarily to ensure the proper functioning of a self-regulating market (Bonal, 2002; Gore, 2000). In terms of social policy, structural adjustment usually means welfare state retrenchment in order to prevent fiscal deficits while privatising welfare state services. The assumption was simply that economic growth would trickle down to all parts of the population and reduce poverty, so that social services could then be afforded on a private market. This means that the period of structural adjustment advocacy came with rather limited relevance to social issues (Vetterlein, 2007). As structural adjustment programmes began to be harshly criticised (starting at the end of the 1980s; ibid.) by internal and external groups due to their detrimental social effects that exacerbated poverty rather than preventing it, the World Bank developed a stronger focus on social protection in terms of social funds and safety nets. In order to regain legitimacy as *the* organisation to fight poverty, the Bank's fourth era is again about putting poverty at the centre of development. Initially, it had a rather one-dimensional understanding of poverty that focused on income-earning and enabling individuals to be integrated into economic processes. This came along with a positive assessment of social investment as crucial in order to drive economic growth. Within the social investment perspective, the World Bank has had a particular focus on investing in children to escape the intergenerational cycle of poverty (Mahon, 2019). Nevertheless, also within that same perspective, structural adjustment still appears to be quite important:

> Investment in people, by themselves, will not be fully effective unless the overall economic policy framework for these investments is conducive [...]. This implies macroeconomic stability, an open economy, access to world markets, the right structure of incentives, and the proper functioning of capital and labor markets. (World Bank, 1995, p. VI)

Another key aspect of the World Bank's social investment perspective is the notion that social expenditure should focus exclusively on the very poor, while subsidies benefiting the elites should be eliminated. The assumption is that elites and the middle class should rely on private provisions for social services, such as education. However, this approach overlooks a critical point: welfare systems that target only the needy are politically more vulnerable to cutbacks. This is because the middle class, which do not benefit from such programmes, are less likely to support them (Mahon, 2019, p. 127). Rianne Mahon, accordingly, comes to the conclusion that "the Bank's social investment discourse held to the neoliberal thrust of the original Washington Consensus while conceding a role for the 'right kind' of social policy" (ibid.).

Nevertheless, developing this thinking further, the World Bank went on to perceive poverty as a multi-dimensional concept that has multiple structural causes and consequences. As the World Development Report (WDR) 2000/2001 (World Bank, 2001, p. V) states

> The report accepts the now established view of poverty as encompassing not only low income and consumption but also low achievement in education, health, nutrition, and other areas of human development. And based on what people say poverty means to them, it expands this definition to include powerlessness and voicelessness, and vulnerability and fear.

This more critical perspective exists alongside ongoing commitments to economic growth and policy reform that promote structural adjustment. This often creates tensions and inconsistencies, and it was in this setting that social capital ideas emerged (Bebbington et al., 2004).

Interpretations of social capital debates in the World Bank suggest that it is a relatively open concept that is not necessarily internally coherent and does create confusion. This leads to a widespread use of the concept to refer to different social dimensions and developments, often in an inconsistent way. Due to its relatively open and vague meaning, the concept can be used by multiple actors and is therefore compatible with both the neoliberal perspective of the World Bank, largely represented by structural adjustment, and perspectives that are critical of structural adjustment and emphasise social policy's contribution to economic development (ibid., p. 36). Accordingly, the term "social capital" can be used appropriately by both perspectives of the double movement (see Sect. 2.3) and is therefore widely diffused. From the perspective of the first movement, which focuses on establishing a self-regulating market, the concept does not question the relations of production within a capitalist system or bring up issues of redistribution, while it also advances semantics of neoclassical economic theory. From a countermovement perspective, on the other hand, it acknowledges the crucial role of government intervention in terms of correcting market failures that are detrimental to social capital as well as the importance of social investment interventions that nurture social capital.

In a similar vein, Bentley B. Allan highlights that the shifts mentioned above did not necessarily mean a complete abandonment of crucial World Bank operations based on "neoclassical economists, neoclassical knowledge, quantitative procedures and modelling devices" (Allan, 2019, p. 197). The incorporation of multidimensional issues was due to legitimacy pressures from civil society actors, advancing issues such as gender equality and good governance. Broadening the definition of poverty and creating a discourse of "multidimensional growth" (ibid., p. 199) in that sense helped to retain traditional World Bank operations that have a particular focus on creating GDP growth: "Neoclassical theory while evolving, provided a stable basis for growth-based nexuses. Poverty alleviation initiatives for health and education were simply reinterpreted as investments in human capital that would drive growth" (ibid., p. 200).

However, the development of World Bank's social policy conceptions did not stop with social investment and multi-dimensional poverty concepts. Around ten years later, the Bank started to integrate an inclusive growth perspective. This was pushed, for example, by the establishment of the Diagnostic Facility of Shared Growth in 2007. According to the World Bank's thinking, inclusive growth should be mainly concerned with increasing opportunities for both low-income and middle-class groups. Generally, the theme of inequality has become apparent in the World Bank's discourse (Mahon, 2019). For example, the WDR2006 (World Bank, 2005) was produced under the responsibility of Chief Economist François Bourguignon, who was firm in his perspective that reducing inequality would increase growth. Moreover, being part of the Social Development department, Anis Dani launched a *New Frontiers of Social Policy* series, which advanced the perception of perceiving individuals with certain social rights and not solely as economic agents (Moser & Dani, 2008). Under President Jim Yong Kim, the notion of "shared prosperity" became an official objective alongside ending extreme poverty. With this concern, the World Bank expanded its primary target beneficiary population to the bottom 40 per cent and focused on income growth across the general population (Mahon, 2019).[1]

While the review above focused on relatively large-scale global social policy developments, the following part reviews global social policy scholarship concerning specific social policy fields in which the World Bank is active. This includes brief reviews of the well-researched cases of pension policy, labour standards and youth unemployment. These fields are illustrated as examples of how the change of broader global social policy

[1] However, this indicator replaces other ones, such as the Gini-coefficient, which also takes account of the top 10 per cent of the population. This leads to the wealth of the wealthiest and its relationship with other parts of the population's income being obscured (Fukuda-Parr, 2019).

themes within the World Bank unfolds in discussions of more specific social policy fields and how this is similarly shaped by the double movement dynamic. In its pension policy, the World Bank implemented several shifts in its discourse. During the 1980s and 1990s, aligned with structural adjustment programmes, there was a strong emphasis on privatisation, which transferred the responsibility for pension provision to individually fully funded pension accounts, reducing the role of public institutions in providing pensions (Orenstein, 2008). This is particularly discussed in the widely cited document *Averting the Old Age Crisis* (World Bank, 1994). There, the World Bank suggests creating a mandatory private pillar within a three-pillar model. A public pillar is supposed to only provide poverty relief, and a third private one covers voluntary private savings (Heneghan, 2021, p. 121f.). Having the support of strong financial market interests within the World Bank, this approach was widely applied to the emerging market economies of East Europe as well as other developing countries. Martin Heneghan (ibid., p. 123) describes this as the World Bank "seeking to limit the size of developing countries' welfare states before they even existed". However, in the 2000s, an open internal debate emerged, reflecting growing scepticism about the emphasis on the private pillar within pension systems while the global financial crisis led to the abandoning of the three-pillar model in several countries as transition costs (paying the current generation of retirees while also shifting funds to the private pillar of current working people) were too high (see also Orenstein, 2011). This led to a rejection of the *Averting the Old Age Crisis* pension model also by the World Bank. Nowadays, the Bank is moving towards the ILO's conception of stronger governmental provision of social protection also in the field of pension policy. Both organisations jointly chair SPIAC-B, where pensions systems are mainly discussed on the basis of a more progressive notion of social protection (Heneghan, 2021, p. 13).

In the field of labour standards, when the World Bank was primarily focused on promoting structural adjustment, it refrained from advocating for labour standards (Römer et al., 2021). This refusal aligned with the neoliberal argument that labour standards should be reduced to allow the market to operate freely and maximise profits. From that perspective, labour standards would only be beneficial if they were explicitly connected to ensuring productivity but not to decommodifying labour. In the 1990s, the World Bank changed its position. As part of interactions with trade unions, the ILO and civil society organisations, it started to pick up on labour standards and refer to them in a more progressive sense. In the context of a study affirming that the right to association would not negatively impact economic growth, the World Bank also expressed support for the ILO's Core Labour Standards. Generally, in contemporary discourse on labour standards, the World Bank has developed a more cooperative stance in relation to the ILO and refers to its conventions that are part of the Core Labour Standards (Römer et al., 2021).

In the field of youth employment, the World Bank's policy development is similar to that of the OECD (see Sect. 4.1). Before the global financial crisis, the World Bank largely criticised over-regulated wages, pointing, for example, to the fact that developing countries had the lowest youth unemployment levels, creating an association between this and the absence of minimum wages for young people in those countries. Advocacy focused on improving training to integrate young people into the job market (Fergusson, 2021). The World Bank had the perception that young people did not have the needed skills and therefore they would be unemployed. The young were blamed as individuals: "young people who failed to acquire basic skills first time around" (World Bank, 2006, p. 96; Fergusson & Yeates, 2013). There was no reference to potential collective action by employers that could ensure early communication about the skills that they demanded, which could then be developed in schools. Moreover, the Bank often appeared sceptical about employment protection laws as they might discourage employers from taking a risk and hiring young and inexperienced individuals (Fergusson & Yeates, 2013, p. 69). Moreover, the Bank did not take account of the perspective that the labour market itself could be a contributor to youth unemployment. It is basically taken for granted that the labour market is the prime and most efficient unit for organising the allocation of labour. Problems and dysfunctions are ascribed to exogenous factors, such as the deficiencies of individuals (Fergusson & Yeates, 2013). However, in the wake of the global financial crisis, this shifted more towards a holistic understanding of youth unemployment, acknowledging potential structural constraints preventing the labour market from providing full employment. Also, the recognition of potentially profound socio-political ruptures and system instability that might appear due to high unemployment rates created a shift towards stronger conditional support (Fergusson, 2021, p. 37).

Overall, historically, the World Bank was quite actively pushing the first movement (of the double movement) that encourages the formation of a self-regulating market by promoting welfare state retrenchment. This has changed somewhat in the last few years, with the World Bank also increasingly collaborating with the ILO and promoting social protection and public intervention, crucial themes of the countermovement. Generally, this underscores the usefulness of the concept of the double movement, which helps to explain not only the dynamics of eco-social policy (see Sect. 7.1) but also those of global social policy and, as will be shown in the following section, those of global environmental policy too in the World Bank's policy discourse.

5.2 The World Bank in Global Environmental Policy

In order to contextualise the discussion of the World Bank's eco-social policy discourse, this section reviews its involvement in environmental policy. It also suggests that the dynamic of the double movement drives the World Bank's environmental policy-making though it historically favours the movement towards economic liberalisation and (green) self-regulating markets.

The World Bank's environmental policy discourse has multiple dimensions that reflect different ways in which it can potentially influence environmental policy-making (Marschinski & Behrle, 2009). A crucial part of its environmental policy discourse are the environmental safeguards that it introduced in response to criticism that some projects (in particular infrastructure projects) have caused severe ecological damage. Moreover, it lays an emphasis on gathering environmental data in various fields of ecological degradation besides focusing to a certain extent on indicators as well as green accounting. The World Bank can have a direct influence on how certain ecological considerations are valued and measured in environmental policy debates. For example, it provides assistance in operationalising international agreements by delivering methodologies by which certain objectives can be reached. These have included, for example, accounting rules for emission reduction through the use of ecological sinks within the framework of the Kyoto Protocol. Furthermore, the World Bank can have executive influence based on its technical advisory work. It pushes certain policy reforms that shape environmental legal and institutional capacity (ibid.). Moreover, the World Bank requires borrowers to fulfil certain ecological considerations. This entails, for example, carrying out environmental impact assessments for projects especially ones that require integrating green components within larger schemes, for instance by establishing conservation areas within larger development projects. However, one should note that the World Bank also finances traditional energy and extraction projects that increase GHG emissions (ibid., p. 120).

Empirical studies on the ecological impact of the World Bank's policies suggest a detrimental effect on environmental protection, often indicating that its involvement exacerbates environmental degradation by promoting policies that expand the role of private actors focused on economic profit. Based on ethnographic research, Michael Goldman (2001) describes how the World Bank constructed a transnational eco-governmentality by laying down rules as to what appropriate use of a nation-state's ecological resources has to look like. This eco-governmentality is embedded in the tools of conservation and preservation promoted by the World Bank. For instance, within the planning process for hydroelectric dams in the Annamite Mountains of the Mekong River, the World Bank stated that millions of people living in local communities needed to be resettled. It saw an opportunity in this process to construct new subjects and make these individuals accountable for project planning. They were supposed to become an "agro-industrial workforce in the newly irrigated *and* electrified plains and a new population of eco-rational natural resource managers" (Goldman, 2001, p. 504). By doing so, the World Bank labelled this project as an ecological one because it was supposed to "develop" the hill tribes that were thought to be active in illegal logging and slash-and-burn farming. The aim was to create a culture that would make "sites and populations more compatible for these large capital investments, even as these investments evolve to include new ways to improve biodiversity, maintain populations, and the professional class" (ibid., p. 506). This is emphasised by the World Bank's attempt to change property rights to create access for different industries to the forest as well as declaring specific protection areas (Burns & Giessen, 2016; Goldman, 2001). Based on a Northern initiative, Laos implemented a new classificatory system that categorised the forest into different areas, such as watershed management, biodiversity conservation and sustainable timber production. This was done to meet market demands across various sectors, including natural resources, electricity generation and aesthetics. It reflected a broader transfer of rights and access to land and river resources from local communities to the industries of tourism, resource extraction and conservation (Goldman, 2001).

Another well-studied example of the World Bank promoting the role of private actors in environmental policy is its involvement in the development of Argentina's forest policies. In this context, a coalition between private landowners and the national bureaucracy emerged, "advocating national development based primarily on more profitable and productive land use" (Burns & Giessen, 2016, p. 494). Together with the Bank, these actors created a policy environment of pushing for neoliberal reform to use forest development to drive economic growth through the production of soy and timber. Such policies entailed, for example, the dissolution of regulatory agencies to create a social environment that supported exports and user right changes, which led to a policy that allowed people to invest without owning land and without the obligation to plant new trees if they have been cut. The World Bank funding was "conditional on government commitment to sustainable and conservation-oriented forestry" (World Bank, 1991, p. 20). This was linked to the categorisation of different parts of Argentina's forests, which included areas of high biodiversity value where forest management was not permitted, areas of medium conservation value where management could be applied under certain conditions and areas of low biodiversity value where land use could be changed freely. However, while the system included some strongly protected areas, it was generally structured to prioritise economic interests. Agencies had significant autonomy in defining these areas and typically received higher returns through industrial land use. Another case study of World Bank-influenced forest management policy in post-Soviet Armenia suggests similar developments. Across various forest-related legislative measures, including reforms in forest administration and the implementation of forest certification, the World Bank particularly promoted the economic exploitation of forests at the expense of conservation practices. It restricted local communities' access to forests for subsistence use while expanding access for commercial actors. This shift was driven by the push for land privatisation and changes to the Forest Code, moving away from strong conservation and preservation goals towards economic utilisation of the forest (Burns et al., 2017).

Overall, the Bank implemented similar environmental policy programmes across various regions globally. The findings from qualitative case studies are also supported by studies based on quantitative data. Shandra et al. (2011) and Sommer et al. (2017) concluded that countries which underwent structural adjustment and investment lending between 1990 and 2005, and 2001 and 2014, experienced higher rates of deforestation than countries that did not participate in such programmes.

In sum, the World Bank's environmental policy discourse is largely connected to the first movement of the dynamic of the double movement, favouring policies that create a self-regulating market that increases the scope of private actors in terms of their access to natural resources, which largely results in environmental degradation. This, from an environmental policy perspective, suggests a certain path dependency towards the eco-social policy perspective of green economy.

5.3 The World Bank's Eco-Social Policy Discourse: Economic Growth, Green Economy and Poverty

While the previous two sections reviewed the World Bank's environmental and social policy concepts, which have traditionally been discussed separately from an eco-social policy perspective, this section offers an empirical reconstruction of the World Bank's eco-social policy discourse. It uses the typology developed in Chap. 3 as a framework for comparison, providing a summarising reconstruction of the World Bank's policy elements that relate in various ways to this typology. This categorisation of the World Bank's policies from an eco-social perspective is further theorised and discussed in Chap. 7, where the book presents four analytical arguments that enhance our understanding of the broader global governance dynamics of the World Bank's (and the OECD's) eco-social policy discourse(s). The subsections that follow explore the World Bank's policy discourse in relation to its perceptions of green growth, green economy, green Keynesianism, recomposing consumption and production, and post-growth while interconnections and tensions between environmental and social policy are discussed as part of the green Keynesianism understanding. Additionally, in Sect. 5.3.3, the book examines the World Bank's Environmental and Social Framework, a crucial legal document for project planning, through a governmentality lens, showing how the Framework promotes a New Public Management discourse that potentially expands the role of private actors in projects aimed at environmental and social policy integration.

5.3.1 *The World Bank and Green Growth*

This subsection reconstructs how the World Bank perceives green growth. According to the eco-social policy typology in Chap. 3, green growth is the umbrella perspective of the two partly opposing perspectives of green economy and green Keynesianism. This subsection provides a broad reconstruction of the World Bank's understanding of green growth, but those that follow provide a deeper engagement with the co-existence of the two opposing versions of green growth: green economy and green Keynesianism. This co-existence of the antagonistic perspectives is further theorised in Chap. 7.

The World Bank advocates green growth to mitigate the oppositional historical relationship between environmental and social policy in regard to economic growth. "Economic growth and social achievements are widely recognized as complementary, but growth and environmental sustainability are often perceived as antithetical. Greening growth would reconcile the need for ecological sustainability with that for economic growth and social improvement" (World Bank, 2012a, p. 30). In that respect, the Bank states, it is generally conceived within green growth that growth in GDP typically comes with an increase in welfare, prosperity and human well-being, which are all key social policy objectives. In order to promote green growth, the World Bank emphasises the importance of integrating ecological factors into all kinds of decision-making which is supposed to reverse the historical non-consideration of the ecology within economic and development policy decisions. According to this thinking, it would need to be acknowledged that the economy is an inherent part of nature.

> Since economies are embedded in nature, policies to promote economic development should also be beneficial to nature. Our ability to produce valuable goods and services for a growing population is bounded by the fact that we cannot live and operate outside nature [...]. At the heart of the challenge is the need to bring nature into decision making at all levels, to improve our collective ability to use the biosphere's goods and services efficiently while allowing it to regenerate so that such goods and services may be sustained or enhanced over time. (World Bank, 2021a, p. VIII)

This perspective highlights the significance of environmental protection practices for fostering economic growth. "Green policies can contribute to growth [...] and boost a nation's overall wealth. And they help to reduce the damage done by environmental degradation, which is costly for an economy" (World Bank, 2012a, p. 10). On a general level, the World Bank's green growth understanding is defined as follows:

> We argue that what is needed is green growth—that is, growth that is efficient in its use of natural resources, clean in that it minimizes pollution and environmental impacts, and resilient in that it accounts for natural hazards and the role of environmental management and natural capital in preventing physical disasters. And this growth needs to be inclusive. (World Bank, 2012a, p. 2)

Overall, the Bank's green growth understanding confirms that growth is desirable and in danger if there are no collective interventions to remedy or mitigate existing market failures that would protect the ecology, which would eventually also undermine the welfare advancements that come along with economic growth because those advances are after all also dependent on the ecology (World Bank, 2021a, p. 51). This argument is essential for green growth (Sect. 3.2).

Also crucial is the idea of "increasing efficiency" (Sect. 3.2), which is omnipresent in the World Bank's discourse and many other policy ideas connect to it. As the Bank assumes: "Energy efficiency is one of the largest untapped sources of energy, and scaling it up is a critical element of the energy transition. It is often the cleanest and lowest-cost way to expand energy services" (World Bank, 2021b, p. 23). Also, the Bank's project planning centres around the idea of increasing efficiency. As part of its environmental and social safeguard, it states that the objective is to "[c]onserve or rehabilitate biodiversity and natural habitats, and promote the efficient and equitable use of natural resources and ecosystem services" (World Bank, 2017, p. 2). Furthermore, the Bank connects the idea of increasing efficiency to social policy issues, including improvements in the labour market: "Investments in energy efficiency can produce a triple dividend of greater energy savings, fewer emissions, and more jobs" (World Bank, 2010, p. 26). However, it also recognises potential tensions concerning inequalities between household groups that could potentially be exacerbated if low-income groups are not supported during the implementation of measures to increase efficiency:

> Low-income consumers are most sensitive to the higher up-front costs of efficient products. Financial incentives to offset these up-front costs, such as consumer rebates and energy-efficient mortgages, can change consumer

> behavior, increase affordability, and overcome barriers to market entry by new, efficient producers. (ibid., p. 213)

In order to disseminate increased efficiency measures, the World Bank emphasises the role of digital technologies: "Digital technologies offer significant opportunities to improve efficiency; reduce congestion, air pollution, and GHG emissions; and transform how people and goods move around the world" (World Bank, 2021b, p. 32). "Increasing efficiency" is also highlighted in respect to creating low-carbon path dependencies by investing in efficient solutions for long-life infrastructure: "One beneficial feature of the inertia in energy infrastructure is that introducing efficient low-carbon technologies into new infrastructure offers an opportunity to lock in a low-carbon path" (World Bank, 2010, p. 203).

Infrastructure takes a prominent role in World Bank discourse. In fact, the World Bank states that "[g]etting infrastructure 'right' is at the heart of green growth. It is critical because infrastructure choices have long-lived and difficult-to-reverse impacts on the carbon, land, and water intensity of future patterns of development" (World Bank, 2012a, p. 133). However, while it is crucial to briefly note this in reference to the general green growth understanding, it is further explored in Sect. 5.3.4.

Overall, the World Bank promotes a "strong" green growth understanding which does not see any disadvantages for economic growth if ecological protection is integrated into economic decision-making (Sect. 3.2): "The analytical case for green growth is strong: green policies can indeed contribute to economic growth over the short term if they are designed in an appropriate framework" (ibid., p. 29). From that perspective, the World Bank underscores shifting market preferences towards new and more efficient technologies: "There are also opportunities to embrace new sources of growth that come from shifting market preferences, leapfrogging to modern, efficient, and competitive technologies" (World Bank, 2021b, p. 8). Accordingly, the Bank assumes that growth does not necessarily have to be slower if it is green.

> Our patterns of growth are unsustainable and inefficient, but this provides opportunities for greener—not necessarily slower—growth. Policy makers have known for decades how to improve the efficiency of production and consumption, but they have failed to make significant progress in adopting the needed measures and deploying the needed resources. (World Bank, 2012b, p. 19)

In sum, the World Bank presents the case for green growth as "strong", emphasising the social benefits in such a transition, first-mover advantages that are connected to technological innovation, increasing efficiency and infrastructure projects. The next two subsections reconstruct the World Bank's green economy and green Keynesianism understandings, two perspectives that are both following line with these general green growth assumptions.

5.3.2 The World Bank and Financialising Nature: Green Economy

The previous subsection set out more general notions of the World Bank's green growth understanding; this one reconstructs the elements that explicitly reveal a green economy perspective as developed in the eco-social policy typology (Chap. 3). Green economy is a green-growth-driven perspective that aims to increase the scope of private actors and favours their voluntary adoption of self-regulatory practices to the detriment of a regulatory top-down governance approach. In terms of its eco-social potential, green economy is the most modest perspective in the typology in that it simply focuses on greening economic growth to enable traditional growth-dependent welfare state operations to continue. In Chap. 7, I argue that green economy is to a certain degree congruent with the first movement within Polanyi's concept of the double movement.

The World Bank incorporates several general green economy elements into its policy discourse that suggest a strong appreciation of market-based measures as drivers of the low-carbon transition. This includes approval of putting a price on carbon and other ecological services, which is further connected to the creation of new markets. Moreover, the World Bank suggests measures to increase transparency in order to hold actors accountable for their ecological protection practices. Another crucial element is that the World Bank suggests increasing the scope of private actors and is sceptical of too much regulation. As far as finance is concerned, the World Bank promotes climate-conditional finance and greening the financial system, among other things putting forward the idea of a green taxonomy to guide sustainable practices within the financial system.

The two flagship reports by the World Bank that were widely referred to in Sect. 5.3.1—the World Bank's green growth strategy *Inclusive Green Growth. The Pathway to Sustainable Development* (World Bank, 2012a) and the WDR on climate change (World Bank, 2010)—feature prominently here too. However, for the World Bank's understanding of green economy, several other documents are also important, such as the World Bank Group's strategy on environment and climate: *Climate Change Action Plan 2021–2025. Supporting Green, Resilient, and Inclusive Development* (World Bank, 2021b) and *Toward a Green, Clean, and Resilient World for All. A World Bank Group Environment Strategy 2012–2022* (World Bank, 2012b). Furthermore, the World Bank's reports on *Transformative Climate Finance* (World Bank, 2020b), and on *Toolkits for Policymakers to Green the Financial System* (World Bank, 2021e), the latter drafted by the Finance, Competitiveness, and Innovation Global Practice unit, also provide substantial elements that are part of the World Bank's green economy understanding. This is also the case with the World Bank's and UNEP's report *Roadmap for a Sustainable Financial System* (World Bank & UNEP, 2017). This subsection also draws selectively on other documents.

The World Bank has a certain focus on promoting policy tools that centre around the market, which would enable traditional growth-based social policy-making to proceed. This has to do, for example, with valuing the ecology in economic and monetary terms as well as emphasising the private actor's role and potential in promoting ecological protection practices and driving the transition to green growth.

> Much of green growth is about good growth policies—addressing market failures and 'getting the price right' by introducing environmental taxation, pricing environmental externalities (such as carbon pricing), creating tradable property rights, and reducing inappropriate subsidies. These measures are critical for enabling the private sector to undertake needed investments and innovations and for getting consumers to internalize the true costs of their behavior. (World Bank, 2012a, p. 12)

Moreover, the World Bank states as its crucial goal for developing countries

> to create new, sustainable, and green markets [...] that encourage private investments, (ii) expanding access to private capital and green finance; (iii) building climate capital markets; (iv) working with development partners and through capital markets to support finance for adaptation and resilience and finance for biodiversity; and (v) enabling the catalyzation of domestic private capital for climate investment. (World Bank, 2021b, p. 39)

Together with UNEP, the World Bank emphasises that "traditional functions of the financial sector" must not be radically changed but that "these functions can be realigned toward sustainable goals to ensure the growth of a more inclusive and sound financial sector, one that intermediates resources, enables payments, and facilitates risk management with increased efficiency and effectiveness" (World Bank & UNEP, 2017, p. 19). The Bank underscores the importance of the growth of markets and advances "policy frameworks and standards that promote the issuance of green financial products [...], the development of new market platforms [...], or the competitiveness of financial centres" (ibid., p. 13). Also, concerning water management, the World Bank proposes tradeable water rights in order to create water markets (World Bank, 2012a, p. 112). Moreover, it encourages the use of "[n]ot-yet-proven advanced technologies, such as carbon capture and storage [...]. Accelerating their widespread availability and use will require greatly enhanced research, development, and demonstration as well as technology sharing and transfer" (World Bank, 2010, p. 190). This emphasis on uncertain technology advances such as CCS resonates well with green economy, which presupposes a particular appreciation of technological solutions despite the uncertainty of expert knowledge (see Sect. 3.2.1).

Furthermore, the World Bank particularly promotes putting a price on carbon. Carbon pricing is understood "to incorporate climate change costs into economic decision-making" (World Bank, 2021b, p. 11) as the current structure of "global markets and economies [...] means that people do not get paid for doing the right thing with regard to the environment" (World Bank, 2012b, p. 8). Moreover, the World Bank states that carbon pricing should function "to incentivize commercial entities and citizens to reduce emissions and the private sector to co-invest in the key systems transitions, reducing the extent of additional public investment needed" (World Bank, 2021b, p. 11), hinting at a limited financial role for public policies. The World Bank particularly underscores the importance of cap-and-trade programmes and a carbon tax because these mechanisms "leave individual firms and households free to find the least-cost way to meet a climate target" (World Bank, 2010, p. 268). This chimes with the green economy rationale of advancing the most flexible and cost-effective solutions for private actors to transition to low-carbon practices and providing leeway and freedom for them on how to achieve emission reductions (Sect. 3.2.1). Other policy ideas prescribed by the World Bank, which are framed to offer a certain degree of freedom and leeway for actors, are, for example, "[e]conomywide energy-intensity targets, appliance standards, building codes, industry performance targets [...], and fuel efficiency standards [which] are among the most cost-effective measures" (World Bank, 2010, p. 213). Moreover, the Bank highlights possible synergies with social issues if environmental taxes replace certain taxes on labour (World Bank, 2012a, p. 47). It also underscores the possible distributional impacts of carbon pricing, pointing out the need for compensation policies to mitigate potential regressive effects on low-income households (World Bank, 2021b, p. 11), emphasising the theme of acknowledging the social burden of certain climate mitigation policies and providing compensation. Nevertheless, the World Bank acknowledges that carbon-pricing policies cannot be successful on their own:

> Carbon pricing can minimize the economic cost of decarbonization when used in conjunction with public investment (e.g., in infrastructure and targeted incentives for technology and innovation), regulatory changes (e.g., for building norms and urban planning), and in the appropriate enabling environment (such as functioning capital markets). (World Bank, 2021b, p. 11)

This suggests a certain role for government intervention (see the analysis of the World Bank's green Keynesianism understanding in Sect. 5.3.4). The Bank states that, even though carbon prices have been implemented in many world regions, the numbers shifting rapidly towards a low-carbon economy are still falling short of what would be desirable: "While the frameworks for climate action – including for robust carbon pricing policies – are in place in many jurisdictions, there is an urgent need to scale the scope and ambition of these investments" (World Bank, 2021d, p. 7). It suggests, moreover "regulatory or fiscal incentives for steering investment towards green assets, for example, through preferential tax treatment for income from green investments" (World Bank, 2020b, p. 29). Another policy tool that it suggests using as a complement to cap-and-trade schemes or carbon taxes are carbon credits. Here, the report *A Guide to Developing Domestic Carbon Crediting Mechanisms* created under the leadership of the World Bank and with support of Get2C, Carbon Limits, the Greenhouse Gas Management Institute, INFRAS, and the Stockholm Environment Institute is crucial. The report defines carbon crediting as follows:

> Carbon crediting refers to the process of issuing emissions reduction units to project activities in recognition of quantified and verified emissions reductions. These reductions are calculated as the difference between emissions from the project and emissions from a baseline scenario, which represents the scenario assumed to occur in the absence of the crediting mechanism. (World Bank & PMR, 2021, p. 2)

They are designed to offset emissions that are subject to regulation, for instance, through an emission trading system, and to incentivise companies to develop voluntary reduction targets (ibid.). Following this, emission trading systems would need to be designed to permit the use of carbon credits. The World Bank underscores that no double counting would be allowed: "crediting mechanisms must only award emission reduction units to projects that are additional: these activities must lead to emissions reductions that would not have occurred in the absence of the crediting mechanism" (ibid., p. 9). The overarching idea is that industries would be more willing to take ambitious action to bring about a low-carbon transition due to the potentially lower costs of compliance with emission trading schemes (ibid., p. 25). Its advocacy of carbon credits emphasises the World Bank's perception of the importance of market solutions, which resonates well with green economy that aims to provide private actors with advanced possibilities to comply with cap-and-trade schemes.

In another move related to green economy, the World Bank promotes the creation of new markets, especially (but not exclusively) with regard to carbon. It perceives the creation of carbon markets not as a goal but rather as a means to achieve a policy target: "carbon markets are in themselves a policy instrument, depending entirely on policy making and regulation" (World Bank, 2012b, p. 60). This aligns with green economy, which emphasises public governance that focuses on establishing and maintaining the effective operation of market-based systems. The Bank considers carbon markets (together with carbon taxes) to be the most appropriate and cost-effective policy idea for reducing emissions, "as they do not require governments to have detailed knowledge of where and how emissions reductions should be delivered or whether governments or consumers should bear the attached costs" (World Bank, 2020b, p. 43). Moreover, the World Bank argues that potential regressive effects could be prevented by either "providing free emissions allocations to industries where there is a risk of job loss or lack of competitiveness" or by using the revenues of carbon markets "to support the growth of alternative industries and fund the development of re-skilling programs and other enabling conditions that support the just transition" (ibid.). To acknowledge the economic value of biodiversity, the Bank also promotes market creation. These markets could include "nature-based tourism supported by biodiversity, watershed services, and climate-regulating services. [Here, t]he main challenge [...] is to create markets for these services so that they become part of the visible economy and are efficiently provided" (World Bank, 2012a, p. 117). One idea relating to the market creation of biodiversity protection practices is, for example, the provision of payment schemes that reward landowners if they actively engage in practices to protect the sink capacities of forests (World Bank, 2021a, p. XIII). The Bank notes that the fundamental problem is that "[e]nvironmental assets do not have readily identifiable prices" (World Bank, 2012a, p. 52), which often leads to decisions that do not take account of nature's economic value. This failure to account for nature's economic values will "misalign the private and social costs and benefits of the use of nature, often to the point that a negative price tag is placed on nature's goods and services" (World Bank, 2021a, p. 62). Accordingly, the Bank points to the importance of creating markets that take account of the positive externalities that nature creates:

> Land use change and degradation often results from the public good nature of ecosystems and the absence of markets for the (positive) externalities they generate [...]. Payment for ecosystem services mechanisms, voluntary or regulated carbon markets, can mimic a market for these services and internalize the positive impacts of conservation. (ibid., p. 68)

Another focus deriving from the green economy understanding is to promote transparency (see Sect. 3.2.1). Here, the World Bank calls for "increase[d] transparency in the energy sector by requiring service companies to share key information, so that the governments and other stakeholders can make better-informed decisions and assessments about removing subsidies" (World Bank, 2010, p. 212). It also underscores the importance of "embedding market-relevant sustainability information in the financial data ecosystem" (World Bank & UNEP, 2017, p. 79). In that regard, "convergence around defining, tagging, measuring, and reporting green assets" (ibid.) would be key. The World Bank also highlights how transparency can create social pressure from individuals in various roles—as citizens, consumers, stakeholders and investors—when information about companies' performance is made public. This social pressure can encourage companies to adopt more sustainable practices (while rejecting more direct intervention based on regulations): "Information allows citizens or governments to put pressure on businesses – the goal of programs that collect and disseminate information about firms' environmental performance. This approach has been deemed the 'third wave' in environmental regulation, after command-and-control and market-based" (World Bank, 2012a, p. 56). In terms of regulating transparency practices, public authorities would have an important role: "financial regulators can set specific disclosure requirements for financial institutions. Furthermore, conduct or capital market regulators could play an essential role in supporting and enforcing sustainability reporting and labelling practices for listed companies" (World Bank, 2021e, p. 76). In particular, central banks could act as role models. They could disclose climate-related information and set essential guidance for other actors in the market (ibid., p. 71). Asset owners are encouraged to disclose their practices as well: "By imposing disclosure requirements on them, the intention is to drive disclosure and action through the entire investment chain" (PRI & World Bank 2020, p. 20).

One crucial characteristic of the World Bank's discourse, as also emphasised above, is scepticism about too much regulation and explicit references to the importance of mobilising private actors and their financial capabilities. In its green growth strategy, the World Bank states that the development of private business should not be constrained in any sense, and regulations should not act as a tool for "discourag[ing] small business development" (World Bank, 2012a, p. 14). It also advocates a more macro-economic perspective, which is reflected in its structural adjustment history that centres around liberalisation and deregulation:

> In countries where the business environment is not conducive to investment and growth, better economic policies must be the first step. Lessons from trade liberalization are telling: where labor mobility is limited by skills and regulations and where investments in the sectors that benefit from trade liberalization are impaired by inappropriate policies, both workers and the private sector take longer to adjust. (World Bank, 2012a, p. 14)

In general, immobility should be prevented, especially in relation to capital and labour movement. Here too, the World Bank reveals a somewhat sceptical attitude towards social security "[s]imple unemployment insurance tends to hamper reallocation and skill formation" (World Bank, 2012a, p. 101). Also, regulations should not be designed in a way that favours "incumbent firms at the expense of new entrants, thereby reducing the ability of the economy to innovate and grow" (ibid., p. 58).

The World Bank employs several discursive elements that indicate a policy agenda of increasing the scope of private actors. Public finance schemes are emphasised for their capacity to "mobilize private investments in energy efficiency and renewable energy" (ibid., p. 22) and provide certainty in order to encourage the private sector to make long-term investments in green industries (ibid., p. 81). In this context, the Bank emphasises the importance of public–private partnerships: "[i]ncreasing the role of the private sector is critical. Many of the needed investments could benefit from public–private partnerships" (ibid., p. 21). It also applies the concept of blended finance: "Blended finance is the strategic use of public finance for the mobilization of private capital flows towards emerging and frontier markets" (World Bank, 2021e, p. 112). Furthermore, it sees an even more substantial role for the private sector in the transition to green growth policies than in its traditional infrastructure financing: "Yet the need for innovation, efficiency, and 'smart investments' [...] makes the role of the private sector even more critical in green growth policies than it already is in traditional infrastructure finance" (World Bank, 2012a, p. 21). The importance of the private sector is emphasised yet again in relation to market-based types of governance, such as cap-and-trade schemes: "An economywide, market-based mechanism, such as a carbon cap-and-trade program or a carbon tax [...], is essential to unleash robust private sector investment and innovation to achieve deep emission cuts at least costs" (World Bank, 2010, p. 190). Furthermore, the Bank also suggests increasing the involvement of private actors in research and development: "Institutional reforms that give the private sector a greater voice in the governance of research institutions and that reward transfer of knowledge and technology to external clients can also help" (ibid., p. 305). Its green growth strategy in particular calls for "[p]olicies that unleash the power of the private sector" (World Bank, 2012a, p. 19). This goes along, however, with a recognition that the policy agenda needs to go beyond the promotion of market solutions: "Market solutions are essential but additional policy tools are needed" (World Bank, 2010, p. 271); "[T]he private sector's role in solving the climate challenge is crucial, but overplaying it would be unwise" (ibid., p. 330). Public policy is particularly mentioned in relation to the creation of "a business environment conducive to low-carbon investment, including but not limited to an expanded, efficient and well-regulated carbon market" (ibid., p. 278ff.), and public funding would be particularly important when it comes to "overcom[ing] investment barriers [...] and to reach[ing] areas the private sector is likely to neglect" (ibid., p. 280). In terms of adapting to certain climate risks, the World Bank highlights the capacity of private insurance providers. Government support is mainly discussed only "as the insurer of last resort" (ibid., p. 101).

In terms of greening the finance system, the World Bank applies (concessional) climate finance that could support private capital:

> to develop and scale new climate-smart technologies and business models in emerging markets and play an additional foundational role in catalyzation of private capital into developing countries. [...] It can enable opening new markets, investing in projects in IDA [International Development Association] and FCS [fragile and conflict-affected states] countries, and scaling up climate finance for decarbonisation in middle-income countries. (World Bank, 2021b, p. 43)

As part of concessional finance, the World Bank proposes green and sustainable loans:

> Sustainability-linked loans are any type of loan instrument and/or contingent facility (such as bonding lines, guarantee lines or letters of credit), which incentivize the borrower's achievement of ambitious, predetermined sustainability performance objectives. [...] [S]ustainability-linked loans look to improve the borrower's sustainability profile by aligning loan terms to the borrower's performance against relevant predetermined targets [...], providing assurance of the actual benefits achieved by the loan. (World Bank, 2021e, p. 119f.)

When advocating green bonds, the World Bank also proposes certain regulatory measures such as "guidance on (or requirements for) definitions, management and use of bond proceeds, reporting, and incentive measures" (ibid., p. 99).

It highlights several issues within current climate finance practices. One key problem is that climate finance is often allocated to individual projects rather than aimed at systemic policy changes. While single-project funding can yield important results, it does not necessarily "change policies and ecosystems for a more transformative impact" (World Bank, 2020b, p. 2).

Another problem is that much climate finance is "provided as development finance through established multilateral and bilateral DFIs [development finance institution]. As such, its allocation naturally follows existing development mandates which may not fully prioritise climate needs" (ibid.). Furthermore, the Bank encourages the use of climate finance instruments beyond grants that are typically aimed at individual projects. In this context, it states that "[i]nstruments such as policy-based finance, results-based finance, equity finance, and guarantees are under-utilised in current climate finance provision" (ibid.). It also calls for climate finance to provide differentiated support based on a country's income level and climate vulnerability (ibid., p. 3) and, furthermore, discusses several policy ideas to "[l]everage national development banks or other domestic public financial institutions to mobilize capital (with a focus on mobilizing private investment) towards green finance" (World Bank, 2021e, p. 89).

Generally speaking, the World Bank suggests "[i]ncorporat[ing] sustainability considerations into national fiscal frameworks, including a review of the effectiveness of fiscal interventions and subsidies in support of green activities and expenditures in unsustainable activities, including fossil fuel subsidies" (World Bank & UNEP, 2017, p. 17). In relation to central banks, it prescribes "[t]he adoption of sustainable and responsible investment (SRI) practices by central banks" (World Bank, 2021e, p. 69). These practices refer to "a financial objective which aims to address the impact of climate-related and/or ESG-related [environmental, social and governance] risks on the portfolio; and an extra financial SRI objective, which aims to address the impact of the portfolio on the environment and society, alongside financial returns" (ibid.). Here, the Bank suggests applying different interest rates depending "on the extent [to] which a counterparty's lending (relative to a relevant benchmark) is contributing to climate change mitigation and/or the extent to which they are decarbonising their business model" (ibid., p. 70). In terms of monetary policy, the World Bank suggests central banks "hold fewer reserves (bearing zero or low interest) against green loans; and green-targeted refinancing lines, which offer refinancing for commercial banks at preferred terms for specified green segments of the economy" (ibid., p. 71). Moreover, it recommends the use of "minimum and maximum credit quotas, requiring banks to allocate a fixed percentage of their loan portfolio to a specified sector [...] or central bank assistance to development banks, offering central bank finance to development banks which can increase their lending capacity to lend to green projects" (ibid., p. 72).

Like the OECD, the World Bank also suggests using green taxonomies as a guide to greening the finance system. They are particularly discussed in *Developing a National Green Taxonomy* (World Bank, 2020c). According to the Bank's definition, a green taxonomy "is a classification system to identify environmentally sustainable economic activities that substantially contribute to the achievement of international and national climate or environmental goals, while causing no harm to environmental and social objectives" (World Bank, 2021e, p. 82). The purpose of such a taxonomy would be to provide essential information to various stakeholders about which practices qualify as sustainable. In this way, it would inform financial actors about sustainable practices, help standardise disclosure processes, guide the innovation of new products and support the monitoring of sustainable public and private investments. Green taxonomies would primarily centre around the financial system. They are supposed to "support the growth of domestic green financial markets" (World Bank, 2020c, p. 18) and to "identify areas of underinvestment" (ibid.). However, a green taxonomy would not work on its own. In order to have an impact, the World Bank notes that it would need to be connected to certain policy measures that encourage sustainable practices: "These can include lowering capital requirements for sustainable financial products, setting regulatory quotes [...], and lowering refinancing rates, among others, based on solid market research" (ibid., p. 25). Green taxonomies chime well with green economy by not prescribing certain types of behaviour but simply tagging practices as green to encourage their voluntary adoption (see Sect. 3.2.1).

In sum, the World Bank adopts a green economy approach that explicitly includes scepticism towards extensive market intervention, while recognising and expanding the role of private actors as the primary drivers of green growth developments. This approach favours voluntary self-regulation over command-and-control arrangements. However, as discussed in Sect. 3.2, the eco-social potential of green economy is generally limited. This limitation largely stems from the assumption that ecological considerations simply need to be integrated into economically driven (green) growth policies, allowing traditional welfare state operations to continue as before. But there are several potential eco-social pitfalls to this perspective (but see the World Bank's green Keynesianism understanding in Sect. 5.3.4). The next section focuses on the World Bank's Environmental and Social Framework and reconstructs how it potentially resonates with the green economy perspective as well.

5.3.3 *The Environmental and Social Framework and New Public Management*

This subsection focuses on the World Bank's Environmental and Social Framework (ESF) and examines how it advances a New Public Management (NPM) discourse that expands the role of private actors in project planning. This further reflects the green economy perspective (see Sect. 3.2.1). The ESF (World Bank, 2017) is the result of one of the World Bank's largest recent reform efforts and establishes mandatory legal mechanisms for both the Bank and its borrowers regarding project planning. It represents the latest version of the environmental and social safeguards, which were introduced to provide guidelines that are supposed to ensure that infrastructure projects financed by the Bank do not have adverse social or ecological impacts (Di Leva, 2021). Thus, the document plays a key role in the World Bank's eco-social policy discourse, as it outlines various ways in which projects may have ecological and social impacts, and how these can be managed to achieve positive outcomes. To demonstrate how the ESF incorporates an NPM discourse, this subsection begins by briefly introducing the core principles typically associated with NPM. Next, it draws on the concept of governmentality (Sect. 2.2) to explain how NPM practices are implemented and applied to the World Bank's borrowers through the ESF. Finally, to provide empirical support for the argument, it illustrates the discursive content of the ESF.

NPM projects a certain scepticism towards too many rules and regulations within bureaucracies, rejects top-down control ideas and instead focuses on empowering individual and autonomous actors. Moreover, the NPM discourse shows a preference for market mechanisms (in particular where traditional public administrations are concerned) to the disadvantage of bureaucratic regulatory mechanisms. This primarily refers to incentivising individuals to adopt specific behaviours, as well as to reversing the expansion of public government in terms of funding and staffing (Hood, 1991; Osborne & Gaebler, 1993). Another key characteristic is that goals and targets are explicitly defined and that there are certain indicators displaying desired outcomes that are measured on the basis of a quantitative methodology to invoke a kind of objectivity. Accordingly, NPM promotes notions of efficient and effective management, and an outcome-based assessment (Hood, 1991; Osborne & Gaebler, 1993). In other words, the NPM discourse is linked to a neoliberal way of governing, whereby clear standardised outcomes are defined to assess the effectiveness and efficiency of projects. Business and managerial values, including assessments, monitoring and reporting, are celebrated while positively appraising an increase in the scope of private actors (Hood, 1991; Osborne & Gaebler, 1993).

Formulating broad objectives enables actors to make their own decisions about the design and implementation of the means to achieve them while complying with certain assessments, monitoring and reporting methods. Therefore, the role of a government or an IO can be limited to the provision of detailed supervision of a project. It need only observe to what extent the "general objectives are met, enabling [potential] private parties to undertake [projects] in the ways that they see fit" (Turnhout et al., 2014, p. 582). Thus, NPM discourse is connected to neoliberalism in three ways: First, it emphasises managerial principles of efficiency and effectiveness. Second, the application of "these principles to governmental and public institutions [...] results in a retreat of the government and enhances the scope for the involvement of private actors" (ibid.). Third, it secures the autonomy and freedom of the actors involved by empowering them (ibid.). In this way, "'free' individuals and 'private' spaces can be 'ruled' without breaching their formal autonomy" (Miller & Rose, 2008, p. 42). "This illustrates that neoliberalism does not necessarily involve a retreat of the government in favour of private and market actors but, rather, the projection of economic principles on (the art of) government" (Turnhout et al., 2014, p. 582). The NPM discourse is reflected in the establishment of technologies and measures for monitoring, reporting and creating indicators and metrics to obtain the transparent, standardised information essential for a market-based form of governance.

From a governmentality perspective (Sect. 2.2), the mechanisms for assessing, monitoring, reporting, creating indicators and the measuring of transparent standardised information reflect an indirect and diffuse disposition of various techniques and tactics to steer the development of the targeted actor(s) (in this case the World Bank's borrowers towards NPM). In this process, it is the actors themselves "on which government will act either directly through large scale campaigns, or indirectly through techniques that will make [it] possible, without the full awareness of people" (Foucault, 1991, p. 100) to inscribe a certain way of behaviour as part of self-regulating practices. This is done through subtle techniques that the governed actor might not be fully aware of and this speaks to questions of what kind of knowledge is deemed crucial while the focus is on how techniques and other means can shape another actor (Dean, 1999, p. 11ff.). Thus, analysing the mentalities of government involves examining the practical and technical applications of thought. Governmentality encompasses a diversity of thoughts, knowledges, techniques and instruments that propose identities for both the governed and the governing, involving varying degrees of guidance in conducting behaviour (Dean, 1999). How this unfolds as part of the World Bank's ESF is detailed below.

A crucial procedure within the process of self-regulation is writing: writing institutionalises and normalises customs and practices that are discussed and illustrated by means of the written word and represents certain repetitive instructions on how to behave according to the institutionalised text (Miller & Rose, 2008, p. 64): "Making people write things down, and the nature of the things people are made to write down, is itself a kind of government of them, urging them to think about and note certain aspects of their activities according to certain norms" (ibid., p. 81).

Both the rhetoric of NPM and the demand to write things down in assessments, reports and monitoring practices are omnipresent within the ESF. The ESF proposes to use measurements and standardised indicators to facilitate the monitoring of projects. This is already clear within the ESF's introduction on the ten different Environmental and Social Standards (ESS) with which the borrower has to comply:

> The Bank requires that all environmental and social risks and impacts of the project be addressed as part of the environmental and social assessment conducted in accordance with ESS1. ESS2–10 set out the obligations of the Borrower in identifying and addressing environmental and social risks and impacts that may require particular attention. These Standards establish objectives and requirements to avoid, minimize, reduce and mitigate risks and impacts, and where significant residual impacts remain, to compensate for or offset such impacts. (World Bank, 2017, p. X)

This means that the borrower has latitude in the creation and evaluation of a project with regard to the possible risks created by it. These risks need to be assessed, managed and monitored to achieve the standardised outcomes of the ten different ESSs: "The desired outcomes for the project are described in the objectives of each ESS, followed by specific requirements" (World Bank, 2017, p. 15). ESS1 sets out the methodological process for the management of all the World Bank's other ESSs and, by applying NPM procedures, shapes them in the same way:

> ESS1 sets out the Borrower's responsibilities for assessing, managing and monitoring environmental and social risks and impacts associated with each stage of a project supported by the Bank through Investment Project Financing, in order to achieve environmental and social outcomes consistent with the Environmental and Social Standards. (ibid.)

It is not the World Bank that has the responsibility, but the borrower which endorses its autonomy and freedom of action. The processes of evaluation, management and monitoring run through the entire project design: "The Borrower will assess, manage and monitor the environmental and social risks and impacts of the project throughout the project life cycle so as to meet the requirements of the ESSs" (World Bank, 2017, p. 17). Furthermore, the borrower will conduct "reporting on the environmental and social performance of the project" (ibid.). Of particular importance is the Environmental and Social Commitment Plan (ESCP) negotiated between the World Bank and the borrower:

> The ESCP will describe the different management tools that the Borrower will use to develop and implement the agreed measures and actions. These management tools will include, as appropriate, environmental and social management plans, environmental and social management frameworks, operational policies, operational manuals, management systems, procedures, practices and capital investments. All management tools will apply the mitigation hierarchy, and incorporate measures so that the project will meet the requirements of applicable laws and regulations and the ESSs in accordance with the ESCP throughout the project life cycle. [...] The management tools will define desired outcomes in measurable terms [...] to the extent possible, with elements such as targets and performance indicators that can be tracked over defined time periods. (ibid., p. 21)

Regular reporting, along with documenting information, is the key governance mechanism that the World Bank applies to its borrowers.

> The Borrower will provide regular reports as set out in the ESCP (in any event, no less than annually) to the Bank of the results of the monitoring. Such reports will provide an accurate and objective record of project implementation, including compliance with the ESCP and the requirements of the ESSs. [...] The Borrower, and the agencies implementing the project, will designate senior officials to be responsible for reviewing the reports. (ibid., p. 22)

In summary, the NPM discourse is characterised by explicitly defined goals and targets, accompanied by (usually quantitative) indicators that reflect desired outcomes. Furthermore, efficient and effective management are emphasised. NPM discourse promotes a preference for market mechanisms and an expanding role for private actors. From a governmentality perspective, the implementation of measures such as (self-)monitoring, reporting and the creation of indicators reflects an implicit way of governing targeted actors without undermining their perceived autonomy. Expertise plays a crucial role in this context, as it can present certain forms and language constellations as "disinterested truth". In particular, the social sciences are vital in the process of "rendering the world thinkable" (Miller & Rose, 2008, p. 62), which refers to the development of various quantitative measures and performance indicators within the ESCP that address environmental and social protection objectives. The employment of those technologies and rationalities creates an art of governing that employs "powerful means of governing at a distance: these appeared to enhance the autonomy of zones, persons, entities, but enwrapped them in new forms of regulations" (Miller & Rose, 2008, p. 18) to which the borrowers have to respond. Accordingly, the World Bank, as an authoritative agent, can apply governmental technologies in the shape of measures and programmes of calculations and procedures upon other actors to teach its objectives (ibid., p. 55). The language and knowledge employed in these measures reflect a certain worldview. Within the ESF, this is the language of the NPM, which entails specific management tools, indicators and an outcome-based approach. While borrowers need to report on their compliance and performance regarding their obligations under the ESF, they internalise the NPM knowledge enshrined in it.

5.3.4 *A Steering Government in the World Bank's Eco-Social Policy: Green Keynesianism*

This subsection reconstructs the elements of the World Bank's policy discourse that explicitly relate to the green Keynesianism perspective as discussed in the eco-social policy typology (Chap. 3). Green Keynesianism, like green economy, is a green-growth-driven eco-social policy perspective, but unlike green economy, it aims to increase the scope of government intervention in order to regulate and control market developments and has more awareness of the ecological consequences of certain welfare state operations as well as the potential negative social effects of certain environmental policy constellations. The simultaneous existence of the (partly antagonistic) green economy and green Keynesianism perspective is further explored in Chap. 7.

In the World Bank's policy discourse, there are some crucial elements that reflect a green Keynesian understanding. These relate to exploring synergies and tensions between environmental and social policy and employing the concept of a just transition. In addition, they emphasise the need for a publicly controlled transition when abandoning high-carbon path dependencies and shifting the use of public finances to low-carbon efforts (changing subsidies). This also includes the idea of a just transition and to a certain degree favours a steering government within an economic framework. Furthermore, like the OECD, the World Bank accentuates the concept of a circular economy.

There are several documents that refer to elements of green Keynesianism. Important discursive elements are found in the World Bank's green growth strategy *Inclusive Green Growth. The Pathway to Sustainable Development* (World Bank, 2012a) and the WDR on climate change (World Bank, 2010). The report *Transformative Climate Finance* (World Bank, 2020b) also puts forward a number of ideas typical of green Keynesianism. In this subsection I also draw on several other documents that deal with relatively specific aspects of green Keynesian ideas.

One crucial characteristic of green Keynesianism (particular in comparison with green economy) is that it provides a certain awareness of the potentially detrimental or synergistic relationship between environmental and social policy. The World Bank subscribes to different interconnections and synergies between environmental and social policy. For example, it stresses the value of environmental policies on social welfare and that the proper design of such policies is also crucial because advances in social welfare could otherwise be jeopardised:

> Green policies can also contribute to welfare through direct environmental benefits, through distributional effects (including poverty reduction and job creation), and through increased resilience to shocks (including natural disasters and commodity price volatility). Welfare impacts will be greater if efforts are made to make green policies inclusive. (World Bank, 2012a, p. 29)

Furthermore, the World Bank highlights fiscal tensions between policy areas, but not in relation to social and environmental policy. Instead, owing to the World Bank's focus on development, tension between development policy and environmental or climate policy does merit discussion (World Bank, 2010, p. 1). In regard to social policy, the World Bank emphasises the importance of implementing social measures to create a resilient population that can cope, to some extent, with increasing climate and environmental pressures. As stated in its WDR on climate change:

> A healthy, well-educated population with access to social protection can better cope with climate shocks and climate change. Social protection policies will need to be strengthened where they exist, developed where they are lacking, and designed so that they can be expanded quickly after a shock. (ibid., p. 13)

In that regard, the World Bank also stresses the diffusion of certain social policy best practices: "Development agencies could help spread successful models of social safety nets and tailor them to the needs created by the changing climate" (ibid.). Generally, the World Bank explicitly represents both policy fields as mutually supporting (World Bank, 2021b, p. 20).

The Bank also acknowledges that certain ecological protection measures (such as carbon taxes) might have regressive impacts on certain parts of the population, in particular low-income households (World Bank, 2010, p. 269). To mitigate the regressive impact, it proposes to use "revenue recycling" (ibid., p. 47) to compensate groups that are disproportionally affected by policies that involve putting a price on carbon. In general, the idea of compensation is quite dominant: "Assessing and addressing the distributional impacts of carbon pricing through the design of carbon pricing instruments and/or complementary policies is critical to enable a socially just transition and to contribute to the long-term sustainability of the carbon pricing mechanism" (World Bank, 2021b, p. 22).

Furthermore, with the report *Shockwaves. Managing the Impacts of Climate Change on Poverty* (Hallegatte et al., 2016), the World Bank accentuates how climate change accelerates poverty. This is explored in relation to various dynamics associated with climate change, such as its impacts on agriculture and food supply, health inequalities, changes in ecosystems and natural disasters, which disproportionately affect poor communities. Generally, the observation is that "[p]oor people are disproportionately affected—not only because they are often more exposed and invariably more vulnerable to climate-related shocks but also because they have fewer resources and receive less support from family, community, the financial system, and even social safety nets to prevent, cope, and adapt" (Hallegatte et al., 2016, p. 2). Therefore, the World Bank mainly suggests increasing the inclusion of poor people particularly in social protection schemes as well as the financial system as assets that are connected to financial institutions would be more secure compared to in-kind assets that can be damaged by natural disasters. Social protection would consist in:

> (i) social safety nets (also known as social assistance), which include conditional and unconditional cash transfers, public work programs, subsidies, and food stamps; (ii) social insurance, which consists of contributory pensions and contributory health insurance; and (iii) labor market measures, which include instruments such as unemployment compensation. (ibid., p. 146)

Generally, this is framed as "pro-poor" climate policies (Hallegatte et al., 2016).

Such an emphasis on compensation also addresses the World Bank's idea of a just transition. "Countries may opt to use industrial policies to support declining industries to facilitate the political economy of a green transition" (World Bank, 2012a, p. 83). Moreover, the Bank emphasises the need for "[l]inking financial support to policy action" (World Bank, 2020b, p. 37) to introduce policies that target specific economic sectors. Promotion of the just transition theme is generally expressed by moving away from supporting certain industries (such as the coal industry) to meet climate targets. Here, the World Bank discusses macroeconomic and fiscal schemes to ensure "the remediation and repurposing of former mining lands and coal-fired power plants" (World Bank, 2021b, p. 24). In short, it applies a just transition perspective to emphasise the burden of accelerating the coal exit on the people employed in related industries and highlights compensation measures to remedy this regressiveness:

> On the supply side, priorities in the transition from coal include accelerating the closure and repurposing of coal mines and coal-fired power plants, with due attention to distributional effects and the promotion of new sources of employment and economic growth for affected people and communities/regions. On the demand side, priorities include reducing, displacing, or avoiding coal use by increasing energy efficiency, switching to low-carbon energy sources, and substantially scaling up renewable energy investments. (World Bank, 2021b, p. 24)

Moreover, to achieve a just transition, the World Bank particularly underscores project financing, which is targeted at specific "technologies and emerging sectors to create new economic activities that benefit communities disadvantaged by climate change" (World Bank, 2020b, p. 26). This speaks to the notion of a "just transition of ecological modernisation" (Kreinin, 2020) as discussed in green Keynesianism (see Sect. 3.2.2).

There are some instances in which the World Bank employs an agenda of a publicly steered shift towards a low-carbon economy as discussed in the green Keynesian perspective (see Sect. 3.2.2). The Bank recognises that "[t]he finite amounts of public climate finance must be deployed in more transformative and catalytic ways to bridge the sizeable gap between available resources and needs" (World Bank, 2020b, p. 1). The government would be supposed to apply "tax reduction, subsidies, or loan guarantees that help overcome barriers to investment in climate mitigation and adaptation" (ibid., p. 32). In that sense, the Bank encourages governments to "use a combination of fees on high emission activities with rebates on low emission activities (known as feebates) to promote mitigation within specific areas, such as supporting shifts from high-emitting vehicles towards low emission vehicles" (ibid., p. 32). It similarly encourages them to use "public funding or tax credits for research and encouraging technology transfer through policy and developing international partnership for research, design and development" (ibid., p. 40). The notion of transformation (see also Sect. 7.3) is also applied to social protection in connection with climate change: "Fundamental transformations are needed in social protection, in carbon finance, in research and development, in energy markets, and in the management of land and water" (World Bank, 2010, p. 60), which is also a crucial connection in green Keynesianism. In terms of mobility, the World Bank advocates a transition to electric mobility, aiming to support "e-mobility solutions, to electrify public transit, green government fleets, adopt micro-mobility solutions, incentivize EV [electronic vehicles], and build the necessary support infrastructure, such as charging stations" (World Bank, 2021b, p. 32). On this basis, the Bank argues that simply enacting short-term investments and fiscal expansion will not be able to trigger the profound changes necessary: "[I]ncorporating sound low-carbon and high-resilience investment components in fiscal expansions to combat the financial crisis will not be enough to thwart the long-term problems posed by climate change" (World Bank, 2010, p. 60).

Following green Keynesianism, the Bank identifies the application of public finances and in particular the notion of changing subsidies as crucial to enacting these more profound shifts. It notes that "[m]any countries channel public subsidies, implicit and explicit, to fossil fuels, distorting investment decisions for clean energy". Accordingly, the problem would be that the "lion's share of the subsidies artificially lowers the price of fossil fuels, providing disincentives to save energy and making clean energy less attractive financially" (ibid., p. 211). In *Transformative Climate Finance*, it states that:

> [f]ossil fuel subsidies encourage the over-use of these energy sources. Their removal can reduce emissions substantially, including by spurring investment in energy efficiency measures or clean energy. Countries may also have opportunities to reform subsidies in other sectors, particularly agriculture, to support climate action. (World Bank, 2020b, p. 32)

The World Bank explicitly provides consultation about reforms designed to change energy subsidies through its Energy Subsidy Reform Facility. Moreover, it states that such policy reforms will be implemented in line with its lending practices. In this context, it acknowledges the potential for regressive social effects: "The Bank will focus on protecting the poor in these reforms by strengthening social safety nets and facilitating communication campaigns to address political economy challenges" (World Bank, 2021b, p. 23). It also underscores the detrimental consequences that these subsidies have on trade and public finance: "These subsidies not only encourage carbon emissions, resource depletion, and environmental degradation, they also distort trade and strain public finance" (World Bank, 2012a, p. 47). Resources would need to be transferred "from fuel subsidies to spending on education, health, and infrastructure [which would] help reach environmental objectives and increase economic growth. Reallocating these funds to services that are accessible to the poor [would] also help reduce poverty" (ibid., p. 159). This means that the World Bank follows the green Keynesian idea of not investing public money in sectors characterised by resource-intensive consumption and production but investing instead in industries and social sectors characterised by low resource throughput (see Sect. 3.2.2). In the field of agricultural subsidies (World Bank, 2021a), an appropriate change, according to the World Bank, would be to decouple subsidies from the volume of production "so that farmers receive the income even when they conserve the forest rather than converting it to grow crops" (ibid., p. XIII). Potential savings from such a reform could be used to finance more agricultural research and development, a carbon payment scheme or also other social policy fields, including education (ibid., p. 67). In that sense, the Bank constructs an entrepreneurial state which is supposed to invest public money in the innovation and development of green industries and technologies (Sect. 3.2.2; Goldstein & Tyfield, 2017).

This kind of investment is explicitly further explored: "public finance and domestic resources, including state-owned banks, will be critical to financing global public goods, such as adaptation, mitigation and a just transition" (World Bank, 2021b, p. 39). Public finance would also be important because there are cases in which the private sector simply "underinvest[s] in projects that provide climate benefits because they will not be recompensed for all the benefits that projects provide to society" (World Bank, 2020b, p. 26). Moreover, public financing of projects would also be crucial to ensure that negative lock-in effects are prevented, something the private sector might overlook (World Bank, 2020b, p. 26). In that connection, the World Bank follows green Keynesianism in calling for public investment in ecological protection technology that the private sector might not invest in as it comes with too much uncertainty (see Sect. 3.2.2). The public could control and collectivise these risks by investing in low-carbon industries. In terms of pushing governments towards green practices, the World Bank urges them to take green considerations into account in budgetary decision-making, because the budget process would be "the central means of ensuring that expenditures are aligned with policy goals and that proper consideration is given to the trade-offs involved when climate-related concerns and growth objective clash" (World Bank, 2012a, p. 155). The Bank also aims to counsel ministries of finance to raise their awareness of climate change and what impact it could have on fiscal planning in the long-term: "It will do so by adapting standard analytic processes, such as macroeconomic modelling, debt sustainability analysis, poverty diagnostics, public expenditure reviews, and public procurement guidelines, to encompass climate concerns" (World Bank, 2019, p. 14). On the other hand, it states that "[p]ublic spending funding alone cannot meet the financing needs to achieve climate and environmental objectives" (World Bank, 2021e, p. 94). Hence, public spending needs also to be concentrated on enabling increases in private financing. So far as countercyclical spending is concerned, the World Bank advocates an approach in which "targeted government stimulus spending must be backed up by accelerated measures to build financial markets that can deliver for people and the planet" (PRI & World Bank, 2020, p. 8).

Generally, there are various occasions and contexts in which the Bank stresses the importance of the government's role within policy-making and an economic framework. For example, it acknowledges that there are various developments, such as the global financial crisis or the rise of China with its comprehensive "nation-state capitalism", which make it reasonable to perceive government intervention as beneficial within an (economic) policy framework:

> Over the past 30 years the role of the state has been cut back in various domains key to addressing the climate challenge, such as energy research. The retreat from direct intervention occurred with a switch from "government" to "governance" and an emphasis on the state's role in steering and enabling the private sector. This general trend hides a complex picture. Twentieth-century Europe saw various forms and degrees of state capitalism. The rise of East Asian economies, including China's, demonstrated the preeminence of the state in "governing the market" to deliver the most successful example of accelerated development. Most recently, the 2008 financial crisis showed the pitfalls of deregulation and unrestrained markets—and triggered renewed emphasis on bringing back the state. (World Bank, 2010, p. 354)

Furthermore, the World Bank perceives that "[t]he current crisis [global financial crisis] has eroded faith in unregulated markets. As a consequence, better regulation, more intervention, and greater government accountability are expected" (World Bank, 2010, p. 61). This is, for example, reflected in the application of Border Carbon Adjustments (BCAs) as a trade policy intervention to integrate climate change considerations in trade policies:

> BCAs place an import tariff on emissions-producing goods from countries with lower carbon prices. This both reduces the risk of competitiveness concerns in the country with the higher carbon price, and, in the country with the lower carbon price, encourages the export of less emissions-intensive goods to avoid the BCA. They can also incentivize climate action to capture carbon revenues domestically rather than facing tariffs. (World Bank, 2020b, p. 38)

BCAs are designed to create a level playing field between imports and domestic production within a carbon border jurisdiction. To enhance acceptance of these measures, the World Bank suggests the possibility of applying revenue transfers to the exporters' jurisdictions (World Bank, 2012a, p. 83). The World Bank also appreciates "strong public policy" (World Bank, 2010, p. 60) because it "can help economies absorb the shocks of unavailable climate impacts, minimize net social losses, and protect the welfare of those who most stand to lose" (ibid.). Furthermore, the government is put at the centre of controlling policy developments and transitions:

> [A]s the prime repositories of political legitimacy, governments will be expected to steer the private sector, facilitate community action, and establish the optimal decentralization of adaptation and mitigation decision making and action. On top of steering, governments will be expected to play an "ensuring" function: guaranteeing that targets and goals are achieved through new emphasis on regulation, taxation, long-term planning, and communication. (World Bank, 2010, p. 331)

Steering the private sector is mainly included in the notion of supporting green sectoral interventions and industries:

> Green sectoral interventions can help increase factors of production, push out the production frontier, enhance efficiency, improve resilience, create jobs, and reduce poverty. [...] Investments in public transit and changes in land-use plans to favor a more compact urban area could reduce air pollution and spur growth. (World Bank, 2012a, p. 161)

Another crucial aspect is public infrastructure design. The World Bank calls for an approach that "requires that almost all new infrastructure be designed with climate change in mind and that urgent action be taken on long-lived infrastructure" (ibid., p. 33). In that sense, the World Bank advocates using "environmental considerations as an entry point to infrastructure project design, generating co-benefits between infrastructure and environment in support of 'green growth'" (World Bank, 2012b, p. 99). In general, the Bank assumes that infrastructure has a significant potential in being synergetic with economic growth and ecological protection: "Infrastructure systems are indeed designed to provide welfare-improving and productivity-enhancing services, which are critical for development, but they also often provide environmental benefits" (World Bank, 2012a, p. 137). The Bank recommends several policy approaches for governments to consider when creating green infrastructure. These include incorporating a "sustainable landscape vision into strategies and policies", harmonising "sectoral plans to incorporate multiple goals for harnessing natural systems", creating "incentives for local actors to participate through policy and public finance" and requiring "decision-makers to consider green infrastructure options in planning processes" (World Bank & World Resource Institute, 2019, p. 77). Also, as regards the creation of industrial park infrastructure, the World Bank highlights the government's scope for using "structural measures to improve resilience when developing and retrofitting industrial park infrastructure" (World Bank, 2020a, p. 84).

In terms of industrial policy, the World Bank favours the idea of supporting and targeting specific industries within an economic framework: "Green industrial policies are policies aiming to green the productive structure of the economy by targeting specific industries or firms. They include industry-specific research and development subsidies, capital subsidies, and tax-breaks; feed-in tariffs; and import protection" (World Bank, 2012a, p. 68). It asserts that, when supporting certain industries, it is crucial to focus on "(1) the development and dissemination of technologies and innovations, by correcting the effect of a knowledge spillover, and (2) the development of new industries and sectors, by correcting the effect of nonenvironmental market failures" (ibid., p. 160). While advocating green industrial policies, the Bank states that benefits need not even necessarily outweigh conventional policies in terms of economic growth and job creation (as long as they do not decrease them) because they would be beneficial in terms of ecological protection (ibid., p. 80). Moreover, it highlights the potential necessity of making green industrial policies permanent, as they also address enduring market failures. Still, there remains some reluctance to fully embrace permanent green industrial policies, as the Bank, particularly in its green-growth strategy, tends to favour "us[ing] price-based instruments to correct permanent environmental externalities and industrial policies to cope with transient externalities" (ibid., p. 84).

The World Bank is also involved in promoting a transition to a circular economy as presented in its *Climate Action Plan. 2021–2025 Supporting Green, Resilient, and Inclusive Development* (World Bank, 2021b, p. 30). In this context, the World Bank discusses issues such as waste management at the municipal level as well as practices of recycling and energy conservation and capacities to cope with hazardous waste (ibid., p. 30f.). In relation to industry, the World Bank lists several policy objectives as part of a circular economy approach, labelling them broadly as "redesign, reduce, reuse, and recycle products" (ibid., p. 34). This would include "energy and resource efficiency, use of renewable energy, included distributed generation and both product and manufacturing process related innovations" (ibid.). By applying a circular economy lens, the Bank follows the green Keynesian perception that waste incineration reproduces reliance on the extraction of new raw materials (which should be abandoned) since used material is not recycled but combusted in a less sustainable way to produce new energy (see Sect. 3.2.2). According to the report *Circular Economy in Industrial Parks: Technologies for Competitiveness* (World Bank, 2021c), produced by the Global Markets and Technology Unit as part of the Green Competitiveness Program, the circular economy is particularly important for enhancing competitiveness in the construction of industrial parks (ibid., p. 16). Moreover, the World Bank promotes eco-industrial parks with certain technological advances and business models that "are important building blocks for scaling up the circular economy approach and accelerating green, sustainable, and resilient industry growth" (ibid., p. 1). The use of renewable energies is particularly promoted within these parks. Furthermore, the World Bank propagates the notion of an "industrial symbiosis network", which is described as "systematic linkages, which create mutual interdependencies between or among firms for their higher production efficiency and business profits" (ibid., p. 109).

In sum, next to adopting a green economy perspective, the World Bank also pushes for policy prescriptions that actually increase the scope of government intervention and acknowledge that private actors need to be controlled to a certain degree. This entails some awareness of potential synergies and trade-offs between environmental and social policy, with a specific focus on poverty, but rejects certain green economy elements that the World Bank also promotes. Chap. 7 makes sense of these contradictions based on the concept of the double movement (Chap. 2).

5.3.5 The World Bank's Understanding of Recomposing Consumption and Production

Here I reconstruct the World Bank's understanding of recomposing consumption and production. Based on the eco-social policy typology (Chap. 3), it summarises and structures the way the World Bank approaches that particular eco-social policy perspective. Generally, the Bank acknowledges the need to change and redistribute consumption patterns as an addition to green growth policies, which alone would not be sufficient and inclusive enough. This involves measures to change social norms, for example, with the help of sustainable public procurement policies and policies that focus on behavioural change based on education, economic instruments, nudging and certification and labelling. The latter is supposed to impact supply chains as well. Other issues include public services in the field of transportation and extended producer responsibility. While the World Bank acknowledges some role for community-driven consumption and government intervention practices, most aspects of the recomposing consumption and production framework align closely with the green economy logic, which does not advocate profound interventions in market operations. Instead, it values the self-regulating market dynamic and emphasises the importance of protecting consumer sovereignty. Consumers are assigned a crucial role in steering the market and are accordingly equipped with individual responsibility to the detriment of collective responsibility.

For the World Bank's understanding of recomposing consumption and production, the WDR on climate change (World Bank, 2010) and the World Bank's green growth strategy *Inclusive Green Growth. The Pathway to Sustainable Development* (World Bank, 2012a) are essential. The *Climate Change Action Plan 2021–2025. Supporting Green Resilient, and Inclusive Development* (World Bank, 2021b) also contains substantial recomposing consumption and production elements. The analysis draws on several other documents as well to illustrate the wide range of documentation that speaks to recomposing consumption and production.

The World Bank highlights various policy ideas, such as changing social norms, regulatory reforms and financial incentives when promoting sustainable consumption and production patterns (World Bank, 2010, p. 15). As discussed in Sect. 3.3, it acknowledges that policy makers need to go beyond implementing economic incentives. They "cannot induce all of the changes needed to protect the environment, given market failures, behavioural biases, and political economy considerations" (World Bank, 2012a, p. 45). Hence, it is also crucial that actors are provided with information on sustainable practices, which is expanded by changing also social norms and regulations (ibid.). Incentives, for example, would involve the reuse of materials as promoted in the concept of a circular economy (World Bank, 2021c, p. 143). Furthermore, the Bank states that simply pursuing green growth that focuses on "increasing efficiency" and technological progress would not suffice to be socially and ecologically just: "It requires complex changes in behaviors and social norms because, even with efficiency gains and new technology, it is unlikely that middle-class consumers (whether in rich or in poor countries) can stick to current consumption patterns" (World Bank, 2012a, p. 15). Accordingly, the World Bank follows the general recomposing consumption and production understanding by noting that existing consumption and production patterns neglect social and ecological protection considerations and that therefore those patterns would need to be changed (Sect. 3.3).

A crucial element in these processes would be to change social norms: "Modifications in social norms that reward a low-carbon lifestyle could prove a powerful element of success" (World Bank, 2010, p. 60). This also speaks to the application of specific cultural frames to increase the social appropriateness and desirability of certain ecological protection practices:

> By priming or framing personal behavior as part of a larger social goal, the public and private sectors can induce people to behave in more environment-friendly ways [...]. By framing environmental protection as a 'social project,' policy makers can make individuals think in terms of social and collective goals. (World Bank, 2012a, p. 55)

In relation to changing social norms, the World Bank also constructs the government as a role model and considers public policies as essential for displaying publicly desirable practices:

> Policy makers can set social norms by changing the behavior of government. The greening of government can play an important communication role in addition to providing immediate benefits in reducing emissions and catalyzing research and investments in new technologies. Where feasible, government can also revise instruments such as public procurement to support green objectives. (World Bank, 2010, p. 340)

Nevertheless, such steering governmental interventions are mainly discussed in the context of supporting and reinforcing specific market dynamics. Here, public procurement policies are primarily constructed as tools to enable innovation and increase economic competitiveness in certain technological areas:

> Governments can take advantage of standard-raising demonstration effects and the provision of a guaranteed demand to foster markets of green products, change technological standards, generate green jobs, adapt public assets (such as buildings and infrastructure), and take a lead in educating consumers and firms to engage in more sustainable consumption and production. (World Bank, 2012a, p. 81)

In order to formalise sustainable public procurement practices, the World Bank suggests establishing "minimum sustainability guidelines in government procurement at the national level. Governments could use their substantial purchasing power to encourage suppliers to comply with labor and environmental standards" (World Bank et al., 2017, p. 49). In other words, as far as public procurement policies are concerned, the World Bank follows the recomposing consumption and production understanding set out in Sect. 3.2 that highlights the government's implicit and symbolic power to steer consumption and production processes and shape consumer culture. In that process, public procurement policies are essential to signal what practices and values are desired, but this subject is discussed in the context of supporting and reinforcing certain market dynamics.

To promote behavioural change towards sustainable consumption and practices, the World Bank supports education campaigns that primarily focus on consumers and the dissemination of specific labels, enabling them to make educated and informed decisions (World Bank, 2010, p. 216f.). Furthermore, the Bank acknowledges that information must be disseminated in a way that accounts for its social context (World Bank, 2012a, p. 52). This reflects a recomposing consumption and production approach, which assumes that individuals, when provided with adequate information about the positive or negative consequences of specific practices, will adjust their behaviour accordingly. To drive such changes, the World Bank also emphasises the potential role of nudging, particularly in making greener default options more appealing: "An important behavioral bias that environmental policy makers can use to their advantage is the tendency of people to stick with the default option" (ibid., p. 54); "To count as a nudge, the intervention must be easy and cheap, but not constitute a mandate. Changing the default options – without changing the options themselves – can be an efficient way to promote greener behaviors" (ibid., p. 19).

Closely connected to these ideas of providing consumers with decent information is the World Bank's prescription of certification and labelling measures in different areas. In the report *Circular Economy in Industrial Parks: Technologies for Competitiveness* created by the Global Markets and Technology Unit as part of the Green Competitive Program, the World Bank emphasises, within the field of industrial parks, the importance of "[i]ntegrated product standards" including "[v]oluntary or mandatory measures including economic instruments, substance bans, voluntary agreements, environmental labelling, and product design guidelines and standards" (World Bank, 2021c, p. 134). The Bank also promotes voluntary schemes in the area of financial systems: "Promoting transparency and efficiency by improving flows of sustainability information through the financial system through voluntary guidance, labelling schemes, or mandatory requirements" (World Bank & UNEP, 2017, p. 13), while carbon labelling is promoted as "an instrument that enables consumers to exercise their desire to join the battle against climate change by using their purchasing preference" (World Bank, 2010, p. 253). Also, as the report *Financing for SMEs in Sustainable Global Value Chains* suggests, the World Bank encourages voluntary third-party labelling and certification systems that integrate certain social and ecological considerations (World Bank et al., 2017, p. 16f.) in the field of sustainable value chains, where certification can have trickle-down effects on some small and medium-sized enterprises (SMEs) that supply larger companies with their products. Even so: "Governments and business could do more to help SMEs improve their sustainability practices. Relatively few companies currently offer effective financing for this purpose" (World Bank et al., 2017, p. VI).

In this respect, the World Bank follows the understanding of recomposing consumption and production (Sect. 3.3) that emphasises its potential to influence entire value chains and establish new global standards. Environmental and social regulations aimed at greening supply chains would encourage companies to implement new standards, which can affect the environmental and social practices of their suppliers often located in distant regions not subject to the original regulations. This aligns with the rationale of reducing GHG emissions on the basis of CBA of GHG emissions (see Sect. 3.3.1). However, while this aligns with the rationale, the Bank does not explicitly refer to CBA.

Extended producer responsibility, particularly in the field of industrial parks, is another concept promoted by the Bank. This would "shift [...] responsibility for waste management from consumers to authorities, who were the traditional assignees, to the producers of the products" (World Bank, 2021c, p. 134). It is further explored in the report *The Role of Extended Producer Responsibility Schemes for Packaging towards Circular Economies in APEC* [Asia-Pacific Economic Cooperation] (World Bank, 2022). The recomposing consumption and production perspective emphasises that extended producer responsibility could encourage corporations to use resources in their production processes that are easier to recycle and less harmful to the environment and society (see Sect. 3.3.3).

In sum then, the World Bank adopts a recomposing consumption and production understanding that is closely connected to the green economy. As discussed in Chap. 3, recomposing consumption and production offers a potential perspective on the transition from green growth to post-growth. However, the empirical emphasis on consumer sovereignty and the non-regulatory assignment of responsibility to individuals to drive market dynamics suggest that the World Bank's understanding of it largely aligns with the green-growth-driven perspective of the green economy and is not therefore moving towards a more transformative post-growth perspective. However, some remarks on governmental intervention, for example in regard to public procurement, are also present. This indicates that the dynamic of the double movement also shapes the recomposing consumption and production understanding (Sect. 7.1). The World Bank's post-growth understanding is discussed in the following subsection.

5.3.6 *Marginal Elements of Post-Growth*

The World Bank does not explicitly mention post-growth or related terms such as degrowth or agrowth. However, in some passages, the Bank recognises that there are more alternative and radical policy proposals than the ones it promotes itself. In its WDR on climate change, it simply acknowledges that there are "[m]ore radical calls for social norms [that] focus on alternative parameters of progress, such as stressing a shift toward notions of well-being decoupled from consumption" (World Bank, 2010, p. 330), without further commenting on them. This acknowledgement may be linked to post-growth proposals that promote a shift from a materialistic cultural notion of well-being to a participatory and convivial one (Sect. 3.4).

Furthermore, as has been said before, the World Bank regards the idea that GDP is a good indicator for measuring economic growth and development with some scepticism. In spite of any scepticism however, the Bank still adds a positive framing of GDP as an indicator. As it states in its green growth strategy *Inclusive Green Growth. The Pathway to Sustainable Development*: "Growth–even measured with such an imperfect metric as gross domestic product (GDP)—is now recognized as a critical driver of poverty reduction" (World Bank, 2012a, p. 4). In *The Economic Case for Nature* it writes that "[a]t least since the end of World War II, economics has been mostly concerned with the evolution of GDP—a measure that provides a powerful indicator of an economy's current level of production but that carries little meaning in terms of sustainability" (World Bank, 2021a, p. 9). Elsewhere, though, the Bank acknowledges that GDP does not capture some aspects of development: "Cleaning up the environment also increases human well-being directly, by improving air and water quality and reducing exposure to natural hazards, although these benefits are not necessarily captured by conventional (GDP) statistics" (World Bank, 2012a, p. 34). These observations show that the World Bank does not argue for abandoning GDP as an indicator of economic growth and development. Instead, it promotes notions such as green accounting as complements to GDP:

> Green accounting extends beyond the valuation of natural assets and focuses on a country's stock of natural and other assets (its wealth) rather than on a flow measure like GDP. By doing so, it helps to identify situations in which economic growth does not create wealth (because natural assets are consumed more rapidly than other assets are created) and is not sustainable. (World Bank, 2012a, p. 18)

In terms of allocating certain responsibilities to developed countries, the World Bank emphasises that developed countries need to free up some development space for developing countries:

> A switch to a low-carbon world through technological innovation and complementary institutional reforms has to start with immediate and aggressive action by high-income countries to shrink their unsustainable carbon footprints. That would free some space in the atmospheric commons. (World Bank, 2010, p. 2)

Furthermore, the World Bank states that "drastic action by high-income countries to reduce their own carbon footprint to sustainable levels is essential. This would show leadership, spur innovation, and make it feasible for all to switch to a low-carbon growth path" (World Bank, 2010, p. 21). This idea of decolonising a fair share of atmospheric commons so that low- and middle-income countries can develop in a fair and just way is a crucial part of post-growth (Sect. 3.4). But the World Bank has not called for degrowth in the Global North as a consequence of this: "all economies, particularly developing ones, still need to grow" (World Bank, 2017, p. 1). Instead, it states that "[d]eveloped countries have an important role to play, offering knowledge, innovation, and financial resources to help support developing countries with this approach" (World Bank, 2021b, p. 8).

In fact, there is a tendency to co-opt the framing of post-growth for a green growth narrative. Stéphane Hallegate, lead economist with the Climate Change Group of the World Bank, in an exchange with Jason Hickel, a degrowth advocate, suggests that degrowth would actually be compatible with green growth if we look at its essence, which consists in reducing materials and energy use as this is also a substantial part of green growth: "Indeed, with this definition, degrowth is uncontroversial, as nobody objects to reducing excess throughput. In fact, in this case, degrowth is also indistinguishable from sustainable development or green growth" (Hickel & Hallegatte, 2022, p. 8).

However, prominent policy proposals that are usually discussed in degrowth circles— such as community currencies, non-monetary exchange systems, alternative credit institutions, transformative redistributive policies, such as income and wealth caps, work time reduction, a citizen's income and new ownership patterns (Sect. 3.4)—are absent from the World Bank's discourse. Neither does the World Bank take account of post-development ideas as part of post-growth that also criticise the narrative of "GDP growth against poverty" because this, as it is often practised, would often only benefit the upper classes in developing countries and promote an overall unsustainable lifestyle for only a few in the Global South, while falling short on basic need satisfaction for all parts of the population there (Gerber & Raina, 2018).

References

Allan, B. B. (2019). Paradigm and nexus: Neoclassical economics and the growth imperative in the World Bank, 1948–2000. *Review of International Political Economy, 26*(1), 183–206. https://doi.org/10.1080/09692290.2018.1543719

Bebbington, A., Guggenheim, S., Olson, E., & Woolcock, M. (2004). Exploring social capital debates at the World Bank. *Journal of Development Studies, 40*(5), 33–64. https://doi.org/10.1080/0022038042000218134

Bonal, X. (2002). Plus ça change … The World Bank global education policy and the post-Washington consensus. *International Studies in Sociology of Education, 12*(1), 3–22. https://doi.org/10.1080/09620210200200080

Burns, S. L., & Giessen, L. (2016). Dismantling comprehensive forest bureaucracies: Direct access, the World Bank, agricultural interests, and neoliberal administrative reform of forest policy in Argentina. *Society & Natural Resources, 29*(4), 493–508. https://doi.org/10.1080/08941920.2015.1089608

Burns, S. L., Krott, M., Sayadyan, H., & Giessen, L. (2017). The World Bank improving environmental and natural resource policies: Power, deregulation, and privatization in (post-Soviet). *World Development, 92*, 215–224. https://doi.org/10.1016/j.worlddev.2016.12.030

Dean, M. (1999). *Governmentality. Power and rule in modern society.* Sage Publications.

Di Leva, C. E. (2021). The challenge and promise at the intersection of environmental and social policies: How the World Bank established a policy framework that fully integrates environmental and social concerns. *Global Social Policy, 21*(2), 344–348. https://doi.org/10.1177/14680181211019170

Fergusson, R. (2021). International organizations' involvement in youth unemployment as a global policy field, and the global financial crisis. In D. Niemann, K. Martens, & A. Kaasch (Eds.), *International organizations in global social governance* (pp. 31–56). Palgrave Macmillan.

Fergusson, R., & Yeates, N. (2013). Business, as usual: The policy priorities of the World Bank's discourses on youth unemployment, and the global financial crisis. *Journal of International and Comparative Social Policy, 29*(1), 64–78. https://doi.org/10.1080/21699763.2013.803998

Foucault, M. (1991). Governmentality. In G. Burchell, C. Gordon, & P. Miller (Eds.), *The Foucault effect. Studies in governmentality* (pp. 87–104). Harvester Wheatsheaf.

Fukuda-Parr, S. (2019). Keeping out extreme inequality from the SDG agenda—The politics of indicators. *Global Policy, 10*(S1), 61–69. https://doi.org/10.1111/1758-5899.12602

Gerber, J.-F., & Raina, R. S. (2018). Post-growth in the global South? Some reflections from India and Bhutan. *Ecological Economics, 150*, 353–358. https://doi.org/10.1016/j.ecolecon.2018.02.020

Goldman, M. (2001). Constructing an environmental state: Eco-governmentality and other transnational practices of a "green" World Bank. *Social Problems, 48*(4), 499–523. https://doi.org/10.1525/sp.2001.48.4.499

Goldstein, J., & Tyfield, D. (2017). Green Keynesianism: Bringing the entrepreneurial state back in(to question)? *Science as Culture, 27*(1), 74–97. https://doi.org/10.1080/09505431.2017.1346598

Gore, C. (2000). The rise and fall of the Washington consensus as a paradigm for developing countries. *World Development, 28*(5), 789–804. https://doi.org/10.1016/s0305-750x(99)00160-6

Hall, A. (2007). Social policies in the World Bank. *Global Social Policy, 7*(2), 151–175. https://doi.org/10.1177/1468018107078160

Hallegatte, S., Bangalore, M., Bonzanigo, L., Fay, M., Kane, T., Narloch, U., Rozenberg, J., Treguer, D., & Vogt-Schilb, A. (2016). *Shock waves: Managing the impacts of climate change on poverty. Climate change and development series.* The World Bank.

Heneghan, M. (2021). International organizations and the global social governance of pensions. In K. Martens, D. Niemann, & A. Kaasch (Eds.), *International organizations in global social governance* (pp. 113–135). Palgrave Macmillan.

Hickel, J., & Hallegatte, S. (2022). Can we live within environmental limits and still reduce poverty? Degrowth or decoupling? *Development Policy Review, 40*(1). https://doi.org/10.1111/dpr.12584

Hood, C. (1991). A public management for all seasons? *Public Administration, 69*(1), 3–19. https://doi.org/10.1111/j.1467-9299.1991.tb00779.x

Kapur, D., Lewis, J. P., & Webb, R. (1997). *The World Bank. Its first half century.* Washington DC: The World Bank.

Kreinin, H. (2020). *Typologies of 'just transitions': Towards social-ecological transformation.* Ecological Economic Papers, 35., WU Vienna University of Economics and Business, Vienna.

Mahon, R. (2019). Broadening the social investment agenda: The OECD, the World Bank and inclusive growth. *Global Social Policy, 19*(1–2), 121–138. https://doi.org/10.1177/1468018119826404

Marschinski, R., & Behrle, S. (2009). The World Bank: Making the business case for the environment. In F. Biermann & B. Siebenhüner (Eds.), *Managers of global change: The influence of international environmental bureaucracies* (pp. 101–142). MIT Press.

Miller, P., & Rose, N. (2008). *Governing the present. Administering economic, social and personal life.* Polity Press.

Moser, C., & Dani, A. A. (Eds.). (2008). *Assets, livelihoods and social policy.* The World Bank.

Orenstein, M. A. (2008). *Privatizing pensions.* Princeton University Press.

Orenstein, M. A. (2011). Pension privatization in crisis: Death or rebirth of a global policy trend? *International Social Security Review, 64*(3), 65–80. https://doi.org/10.1111/j.1468-246X.2011.01403.x

Osborne, D., & Gaebler, T. (1993). *Reinventing the government. How the entrepreneurial Spirit is transforming the public sector.* Plume.

Römer, F., Henninger, J., & Le, T. D. (2021). International organizations and global labour standards. In K. Martens, D. Niemann, & A. Kaasch (Eds.), *International organizations in global social governance* (pp. 57–81). Palgrave Macmillan.

Shandra, J. M., Shircliff, E., & London, B. (2011). World Bank lending and deforestation: A cross-national analysis. *International Sociology, 26*(3), 292–314. https://doi.org/10.1177/0268580910392260

Sommer, J. M., Shandra, J. M., & Restivo, M. (2017). The World Bank, contradictory lending, and forests: A cross-national analysis of organized hypocrisy. *International Sociology, 32*(6), 707–730. https://doi.org/10.1177/0268580917722893

Stone, D. (2003). The "Knowledge Bank" and the Global Development Network. *Global Governance: A Review of Multilateralism and International Organizations, 9*(1), 43–61.

Turnhout, E., Neves, K., & de Lijster, E. (2014). "Measurementality" in biodiversity governance: Knowledge, transparency, and the intergovernmental science-policy platform on biodiversity and ecosystem services (Ipbes). *Environment and Planning A: Economy and Space, 46*(3), 581–597. https://doi.org/10.1068/a4629

Vetterlein, A. (2007). Economic growth, poverty reduction, and the role of social policies: The evolution of the World Bank's social development approach. *Global Governance: A Review of Multilateralism and International Organizations, 13*(4), 513–533. https://doi.org/10.1163/19426720-01304006

Vetterlein, A., & Schmidtke, T. (2024). The World Bank: A changing organization in a changing world. In A. Vetterlein & T. Schmidtke (Eds.), *The Elgar companion to the World Bank* (pp. 2–20). Edward Elgar Publishing.

World Bank. (1991). *The forest sector: A World Bank policy paper*. The World Bank.

World Bank. (1994). *Averting the old age crisis. Policies to protect the old and promote growth*. Oxford University Press.

World Bank. (1995). *Investing in people: The World Bank in action*. The World Bank.

World Bank. (2001). *World development report 2000/01: Attacking poverty*. Oxford University Press.

World Bank. (2005). *World development report 2006: Equity and development*. The World Bank.

World Bank. (2006). *World development report 2007: Development and the next generation*. The World Bank.

World Bank. (2010). *World development report 2010. Development and climate change*. The World Bank.

World Bank. (2012a). *Inclusive green growth. The pathway to sustainable development*. The World Bank.

World Bank. (2012b). *Toward a green, clean and resilient world for all. A World Bank Group environment strategy 2012–2022*. The World Bank.

World Bank. (2017). *Environmental and social framework*. The World Bank.

World Bank. (2019). *The World Bank Group action plan on climate change adaptation and resilience*. The World Bank.

World Bank. (2020a). *Resilient industries. Competitiveness in the face of disaster*. The World Bank.

World Bank. (2020b). *Transformative climate finance: A new approach for climate finance to achieve low-carbon resilient development in developing countries*. The World Bank.

World Bank. (2020c). *Developing a national green taxonomy. A World Bank guide*. The World Bank.

World Bank. (2021a). *The economic case for nature. A global earth-economy model to assess development pathways*. The World Bank.
World Bank. (2021b). *Climate change action plan 2021–2025. Supporting green, resilient and inclusive development*. The World Bank.
World Bank. (2021c). *Circular economy in industrial parks: Technologies for competitiveness*. The World Bank.
World Bank. (2021d). *State and trends of carbon pricing 2021*. The World Bank.
World Bank. (2021e). *Toolkits for policymakers to green the financial system*. The World Bank.
World Bank. (2022). *The role of extended producer responsibility scheme for packaging towards circular economies in APEC*. The World Bank.
World Bank & PMR. (2021). *A guide to developing domestic carbon crediting mechanisms*. The World Bank.
World Bank & UNEP. (2017). *Roadmap for a sustainable financial system*. The World Bank.
World Bank, & World Resource Institute. (2019). *Integrating green and gray. Creating next generation infrastructure*. The World Bank.
World Bank, Global Partnership for Financial Inclusion, & German Federal Ministry for Economic Cooperation and Development. (2017). *Financing for SMEs in sustainable global value chains*. The World Bank.

CHAPTER 6

Comparing the OECD and the World Bank: Different Eco-Social Emphases

This chapter compares the empirical reconstructions of the World Bank's and OECD's eco-social policy discourses (Chaps. 4 and 5). Generally speaking, the organisations are characterised by their similarities, but there are some differences too that are depicted below. These differences can be explained by the different policy foci of the two organisations. The OECD provides policy guidance largely to high-income countries; the World Bank guides low- and middle-income countries to develop towards becoming high-income ones. Also, while the OECD relies much more on European networks and needs to produce a consensus among member-states, the World Bank is much more US-centred, with the US having veto power, which also influences the policy output of both organisations (Deacon, 2007; Eitinger & Wade, 2024). This chapter offers mainly a more descriptive comparative summary of the OECD's and the World Bank's eco-social policy discourse based on the previous chapters and, though it includes a remark on one of the main heuristics for theorising the empirical results—the concept of the double movement by Polanyi (2001)—the next chapter (Chap. 7) presents the original theorisation of the empirical findings for this book.

Generally, when discussing the interconnections and tension between environmental and social policy discourses, the World Bank and the OECD refer to a variety of similar themes, such as highlighting the potentially detrimental relationship between economic growth and

R. Schulze Waltrup, *Eco-Social Policy in the OECD and the World Bank*, https://doi.org/10.1007/978-3-031-95429-0_6

environmental protection, the complementary role of economic growth and social policy, the potential regressive effect of climate and environmental policies and proposing a just transition towards a low-carbon economy with various ideas of social compensation. The differences between both organisations are reflected in different foci, such that, generally speaking, the OECD has a somewhat more differentiated and elaborated perspective on tensions and interconnections between environmental and social policy, which also speaks to the emphasis of green Keynesianism within the OECD more generally. For example, the OECD systematically highlights notions of inequality in terms of unequal impacts of environmental degradation and climate change on social groups, and in terms of how environmental protection policies might exacerbate social inequalities, depending on their policy design (OECD, 2021). Furthermore, the OECD's work on potential regressive effects is more differentiated and investigates those effects with respect to different social groups, while the World Bank refers to those potential regressive effects mostly on a rather general level. However, how climate change links to poverty is particularly systematically explored by the World Bank (Hallegatte et al., 2016). Both organisations refer extensively to the notion of a just transition to enable a socially inclusive transition towards a low-carbon economy and different methods of compensating potential "losers" from climate mitigation policies. In short, the crucial difference in terms of interconnections and tensions between environmental and social policy between the OECD and World Bank is that the OECD has a strong focus and a more elaborated perspective on inequality within eco-social policy, while the World Bank explores more systematically the interconnection of climate change and poverty.

Furthermore, the OECD has a somewhat more differentiated perspective on economic growth. While the World Bank does not systematically mention possible negative social consequences due to economic growth, the OECD points out that economic growth would not only lead to environmental degradation but also often to rising inequality (OECD, 2020, p. 15). Quite to the contrary, the World Bank suggests that economic growth would be beneficial for all sections of the population. Here the World Bank refers to the reduction of poverty: "it is generally accepted that growth tends to raise incomes for all members of society, including the poor, and, if sustained, it can generate virtuous cycles of prosperity and opportunity and support progress against multiple dimensions of poverty" (World Bank, 2021, p. 51). However, generally, the World Bank also calls for "inclusive growth" (World Bank, 2012, p. 30). The more differentiated perspective on economic growth becomes particularly clear when we look at the OECD's thinking on GDP. The OECD states that GDP is not a good proxy for well-being and overall economic activity. This is because important aspects of well-being, such as health, distribution of income and the value of personal relationships, are not captured by GDP. While the World Bank occasionally also refers to well-being, the OECD clearly centred around the notion of a well-being framework that is supposed to create an integrated and aligned policy perspective on various policy issues. On an organisational level, this is reflected by the establishment of WISE in particular. Furthermore, within the OECD's discourse, the importance of developing an integrated and aligned policy perspective based on a well-being framework is more systematically explored than in the World Bank's, as reflected in the OECD's well-being and systems-thinking perspective (Hynes et al., 2020; OECD, 2011, 2019). This can largely be explained by the policy foci of the organisations. The OECD concentrates on high-income countries, where there is already a significant amount of GDP and economic growth that needs to be distributed more equitably to enhance prospects for well-being. In contrast, the World Bank focuses on generating GDP in the first place in low-income countries, where advances in GDP are more needed to drive development.[1]

Neither organisation systematically addresses the tension between environmental and social policy that specifically speaks to differing perceptions of urgency. They do not explicitly acknowledge the tension that welfare policies are traditionally seen as more urgent than environmental policies, as immediate social issues like disease or poverty are perceived as more pressing than ecological problems, which are often diffuse and only indirectly perceptible (Sect. 3.1). Furthermore, while the OECD falls short of systematically addressing fiscal tensions between these policy areas, the World Bank does highlight such tensions, but with a focus on development policy (World Bank, 2010, p. 1).

Both organisations have a similar green growth understanding with slightly different emphases (see Sects. 4.3.1 and 5.3.1). Both highlight the importance of incorporating ecological factors and assets into economic decision-making to reverse or negate the historical non-consideration of the ecology within economic policy decisions. Both organisations

[1] Note that the perspective of "development through GDP growth" is also harshly criticised by post-growth and post-development scholars (Gerber & Raina, 2018).

highlight crucial notions of energy efficiency and employ a green growth understanding that sees the potential to increase economic growth through the implementation of ecological protection measures both in the short and the long term. In that regard, both organisations highlight first-mover advantages. Moreover, both refer to the importance of innovation and technology in driving economic growth and to circular economy arguments, both highlight certain social policy effects of green growth, such as potential new jobs in the field of energy efficiency (as this would be quite labour-intensive), and both perceive green growth as crucial for the further development and continuation of social policy. However, within their green growth understandings, there are some different focal points. The World Bank's understanding centres largely on the creation of green infrastructure: "Getting infrastructure 'right' is at the heart of green growth. It is critical because infrastructure choices have long-lived and difficult-to-reverse impacts on the carbon, land, and water intensity of future patterns of development" (World Bank, 2012, p. 133). The OECD acknowledges the importance of infrastructure for green growth, but does not frame it as being "at the heart" of its green growth understanding. Instead, its understanding of green growth relies more on the promotion of a knowledge economy, which is absent from the World Bank's discourse. This can be explained by the policy focus of the respective organisations. The World Bank, since its foundation, has always focused on infrastructure investments in developing countries as well as setting up specific policies around the creation of infrastructure projects. The OECD, on the other hand, focuses on high-income countries that are largely considered to be driving the technological innovation and knowledge generation that are supposed to create "first-mover advantages", crucial ideas within the OECD's concept of the "knowledge economy". Both organisations employ understandings of both green economy and green Keynesianism, despite these perspectives being considered contrary, and to a certain extent antagonistic to each other (see Chap. 3). In Chap. 7 this contradiction is further explained through the concept of the double movement, the argument being that the green economy perspective represents, to some degree, the first movement, which advances self-regulating (green) market practices, while green Keynesianism represents the countermovement, aiming to contain this first movement through principles of government intervention and social protection. This dynamic eventually stabilises the capitalist mode of production (see the following chapter for a detailed analysis). Meanwhile, this present chapter continues by briefly describing the differences in how the two organisations understand the green economy and green Keynesianism without attempting to make sense of the contradictions yet.

Both organisations employ a green economy understanding (see Sects. 4.3.2 and 5.3.2). This prevents market interventions from going beyond correcting market failures and recommends that ecological services and goods should be integrated into economic decision-making based on their monetary value. Moreover, it focuses on supporting private actors as the main agents in steering green growth developments. Accordingly, actors are to be left free to decide on their own how they reduce emissions, and non-obligatory self-regulating risk management practices are particularly encouraged. Both organisations promote the application of economic instruments as a core policy tool since they are perceived to be the most flexible and efficient policy solutions and prominently advocate transparency and the provision of information and disclosure so that actors can freely adapt their behaviour towards more sustainable practices. Both favour private actors for their potential to create innovation and competition, and stress the idea of "blended finance", which entails various ideas for mobilising private capital with the help of public investments and guarantees. Both, moreover, highlight that the existing financial markets need to be shifted towards sustainable practices, and that potential revenues from economic measures could be used for social policy responses. They also both promote the idea of a green taxonomy in which certain practices and investments are "officially" labelled as green or sustainable in order to steer various actors towards them. However, the OECD takes a more multidimensional perspective on taxonomies incorporating social issues more systematically next to ecological ones. Overall, the organisations' green economy understandings come with a rejection of top-down and controlling policies.

While all these characteristics are possessed by both organisations, the World Bank has a more comprehensive understanding of green economy than the OECD and in its discourse green economy is more dominant in comparison to green Keynesianism. The Bank employs several discursive elements that focus more strongly on ensuring that the scope of private actors is not constrained in any way, which is reflected in a policy prescription narrative that is supposed to "unleash the power of the private sector" (World Bank, 2012, p. 19). Moreover, it discusses the role of private insurance providers and concessional climate finance more systematically, which points to a mobilisation of finance and loans in conditions of ecological protection. Generally, the World Bank provides more extensive policy prescriptions in the field of green finance than the OECD, which is unsurprising considering that it explicitly provides finance while the OECD does not. The Bank, for example, proposes carbon credits, which can work as a complement to cap-and-trade schemes. They create the option for actors to offset emissions that are subject to existing emission trading systems. This means that the Bank is working on further development of existing market-based solutions as part of its green economy understanding. This particular focus on empowering the private sector is also reflected in the application of an NPM discourse in the World Bank's (2017) ESF that entails mandatory legal mechanisms for the World Bank and its borrowers in project planning. The ESF shows a preference for market mechanisms and increasing the scope for private actors. From a governmentality perspective, it steers all kinds of actors (regardless whether private or public) towards (self-)monitoring, reporting and the creation of indicators without breaching their formal authority as autonomous actors (see Sect. 5.3.4).

Both organisations employ a green Keynesian understanding (Sects. 4.3.3 and 5.3.4), which entails opposing ideas concerning the role of the government in policy-making compared to green economy. While the World Bank's understanding of green Keynesianism should not be underestimated (especially considering that historically it pushed quite a contrary policy agenda as part of the Washington Consensus (Wade, 1996)), the OECD engages more with various themes from the discourse and promotes a more elaborated perspective of Green Keynesianism. Nevertheless, there are also crucial similarities between the organisations' discourses. Both state that there is a need for a publicly steered transition towards a low-carbon economy. Both propagate the crucial green Keynesian attitude of moving subsidies away from high-carbon practices and technology to low-carbon ones. Shifting subsidies is also largely related to social policy as new resources could be channelled towards social protection. Both organisations advocate a more institutional alignment of climate change objectives with different objectives for different ministries. Both largely refer to the increasing importance of government intervention to steer developments within an economy, following the understanding that market imperfections are not due to state intervention but due to industries' and capital actors' unregulated practices. Both employ discursive elements that highlight the necessity of developing social protection, investments in green infrastructure, industry, technology and innovation, and of abandoning high-carbon path dependencies. Unsurprisingly, having in mind the World Bank's focus on infrastructure within green growth (see above), the World Bank highlights investments in infrastructure more than the OECD. Furthermore, in developing an industrial policy, the World Bank highlights certain policy measures regarding industrial parks or eco-parks, which also play a prominent role in its circular economy understanding. The OECD does not talk that extensively about different developments within industrial policy, but both organisations perceive a steered industrial policy by an "entrepreneurial state" as crucial.

The OECD's understanding of green Keynesianism goes beyond the World Bank's in several instances and can also be considered the more dominant within the OECD in relation to the green economy perspective. Contrary to the World Bank, the OECD (in particular NAEC within the OECD, even though this goes much beyond NAEC) rejects certain monetarist ideas and explicitly states that monetarist policies are insufficient to drive a transition towards a low-carbon economy and highlights the importance of a proactive fiscal policy response in regard to both economic shocks and long-term policy developments. NAEC also, to a certain extent, promotes public borrowing and rejects austerity: "it has been shown how high levels of public borrowing and debt can be sustained so long as the growth rate of the economy (which can itself be stimulated by public investment) exceeds the rate of interest paid. Public investment can 'crowd in', rather than 'crowd out', private finance" (OECD, 2020, p. 21f.). Freed-up resources due to subsidy change could be channelled towards social protection or green technology and innovation. NAEC explicitly calls as well for a "paradigm shift" (OECD, 2020, p. 29) centring around an appreciation of government intervention and a rejection of policy principles enshrined in the Washington Consensus. Washington Consensus prescriptions rejected government intervention and opted to liberalise markets: "policy rooted in this kind of analysis tended to be sceptical of government intervention, with deregulation of various kinds widely favoured" (OECD, 2020, p. 14). To overcome such a policy approach, NAEC proposes an increased appreciation of government intervention which goes beyond correcting market failures. As economic growth is not the primary objective of policy-making but "aims at different ethical outcomes, the state will have to play a larger role in guiding, or steering, the overall patterns of economic activity to achieve them. [...] 'Correcting market failures' will not be sufficient; markets can also be 'shaped' in pursuit of publicly determined goals" (OECD, 2020, p. 23). The observation that the World Bank has a more advanced understanding of green economy and a stronger focus on it while the OECD has a stronger emphasis on green Keynesianism resonates with Deacon's (2007) observation that the OECD has a somewhat more balanced perspective on policy developments that incorporates more mainstream European government intervention policy prescriptions compared to the World Bank, in which (still) the US dominates having a veto-power that favours a policy approach of privatisation and deregulation.

Both organisations have a similar recomposing consumption and production understanding with slightly different emphases (Sects. 4.3.4 and 5.3.5). Both favour an individualistic notion of consumer sovereignty as a starting point for policy considerations and tend to avoid promoting measures that might overcome any cultural and institutional lock-in effects driving environmentally and socially harmful consumption and production (Lorek & Fuchs, 2013; Sanne, 2002), with the potential exception of public procurement policies on the basis of which governments can act as role-models to drive cultural and institutional change. This also includes perceiving the consumer as crucial to stimulating changes in production processes. Even though both organisations state that there is a need to redistribute resource consumption and that high-income countries cannot stick to current consumption and production patterns, they do not prescribe specific policy measures that explicitly limit the consumption of high-income groups and fall short of suggesting certain redistributive measures in that regard. Overall, both organisations' understandings focus on more individualistic incremental changes that do not question the way that markets operate. Policy measures, such as information campaigns, certification, labelling, nudges, education campaigns and economic incentives, are considered crucial by both organisations to "empower" consumers within given market settings and in that sense help them to drive innovation and competition between firms while also stimulating changes in production. Both organisations emphasise the need to change cultural norms around individual consumption decisions, as well as the potential of educational campaigns to drive long-term cultural shifts. They also largely refer to public procurement policies as tools to reinforce existing market dynamics, by, for instance, fostering innovation and creating new markets, and both promote extended producer responsibility. In contrast to the World Bank, the OECD highlights the idea of CBA of GHG emissions and its importance to having a clear understanding of the resources consumed—and not only produced—within national territory. This is important as CBA reveals the persistent high share of GHG emissions from high-income countries that are simply outsourced through outsourcing the production of the goods that are consumed there. This acknowledges implicitly that recomposing consumption and production policies in developed countries could affect the whole supply chain, impacting production processes in developing countries (Sect. 3.3). The World Bank does not pick up on this issue in a systematic manner and therefore fails to embark on a discussion that centres around high-income countries still taking up atmospheric common space with their consumption practices that are largely attributed to developing countries as part of their GHG emission statistics based on territorial counting. Furthermore, within the OECD's policy discourse production-service systems take a more prominent role. Those provisions are usually not as environmentally damaging as individual possessions because they reduce individual ownership, leading to less resource throughput. Such an approach might also limit the impact of "positional goods", usually private individual possessions, with the effect of reducing positional-goods-driven consumption competition (Sect. 3.3.3). Also, the OECD promotes, to a larger extent, the implementation of differentiated tax systems (smart VATs), which represent an incentive to consume environmentally and socially friendly products to the detriment of unsustainable products. Neither organisation refers to PCT schemes, which would distribute carbon allowance and/or consumption allowances at the household and individual level, in a similar way to how cap-and-trade schemes work (see Sect. 3.3.3).

Comparing both organisations' post-growth understandings (Sects. 4.3.5 and 5.3.6), it needs to be emphasised that, as shown above, both organisations employ a green growth understanding with prominent green growth strategies, so neither can be considered a "post-growth-organisation". However, that does not mean that post-growth elements are completely absent. The OECD, as shown above, adopts a significant number of GDP-critical elements in its policy discourse. It states that GDP would not be a good indicator for positive (economic) policy-making and that it would fail to capture important dimensions of well-being. The World Bank is less critical of GDP. If it expresses some scepticism, it also ensures that criticism is connected to a positive framing of GDP: "Growth—even measured with such an imperfect metric as gross domestic product (GDP)— is now recognized as a critical driver of poverty reduction" (World Bank, 2012, p. 4). The Bank also acknowledges that an increase in GDP would lead to an increase in environmental degradation.

However, the OECD goes beyond this by also acknowledging that GDP might increase due to potentially "regrettable" practices. It is also critical of the fact that non-monetary care work would not be captured by GDP, and that GDP falls short in indicating how economic assets are distributed in society (OECD, 2020). In line with post-growth's conception of well-being (Sect. 3.4), the OECD also professes a rather broad and comprehensive understanding of well-being that incorporates issues such as "income, jobs, health, knowledge and skills, safety and the quality of the environment, as well as the economic, natural, human and social capital stocks needed to sustain well-being over time" (OECD, 2019, p. 11). This speaks to well-being issues that are also important within the post-growth discourse such as nurturing relationships with other people, their community and self-development and to some degree can be considered as a move away from a purely materialistic conception of well-being. The World Bank refers less systematically to the notion of well-being overall.

In terms of more specific post-growth policy proposals (Sect. 3.4.3), the OECD suggests the usefulness of work time reduction policies, job guarantees and universal basic services (though rather at the margin of its discourse) but does not discuss any other policy ideas, such as local currencies, alternative monetary exchange services, policies that support communal ownership patterns or comprehensive redistributive policy ideas such as a universal basic income or income and wealth caps. The World Bank does not discuss any typical post-growth policy proposals.

Post-growth can be considered as largely absent due to the fact that both organisations are, at the end of the day, intergovernmental: national governments determine the dominating themes and the overarching policy direction. Member-states that largely pursue growth policies and objectives would simply not agree to a post-growth policy agenda if pushed by the organisations' staff. In the OECD a consensus among all member-states would be needed, while in the World Bank at least the US (together with enough other member-states to represent a critical mass) would have to agree to a post-growth agenda due to its veto power. This seems rather unrealistic when looking at policy statements of representatives of the respective member-states that selectively focus on "green", "clean" and/or "inclusive" economic growth.

References

Deacon, B. (2007). *Global social policy & governance*. Sage.

Eitinger, L., & Wade, R. H. (2024). US 'hegemony'in the World Bank. In A. Vetterlein & T. Schmidtke (Eds.), *The Elgar companion to the World Bank* (pp. 118–128). Edward Elgar Publishing.

Gerber, J.-F., & Raina, R. S. (2018). Post-growth in the global South? Some reflections from India and Bhutan. *Ecological Economics, 150*, 353–358. https://doi.org/10.1016/j.ecolecon.2018.02.020

Hallegatte, S., Bangalore, M., Bonzanigo, L., Fay, M., Kane, T., Narloch, U., Rozenberg, J., Treguer, D., & Vogt-Schilb, A. (2016). *Shock waves: Managing the impacts of climate change on poverty. Climate change and development series.* The World Bank.

Hynes, W., Lees, M., & Müller, J. M. (Eds.). (2020). *Systemic thinking for policy making. The potential of systems analysis for addressing global policy challenges in the 21st century.* OECD Publishing.

Lorek, S., & Fuchs, D. (2013). Strong sustainable consumption governance—Precondition for a degrowth path? *Journal of Cleaner Production, 38*, 36–43. https://doi.org/10.1016/j.jclepro.2011.08.008

OECD. (2011). *How's life? Measuring well-being.* OECD Publishing.

OECD. (2019). *Accelerating climate action: Refocusing policies through a well-being lens.* OECD Publishing.

OECD. (2020). *Beyond growth: New approaches to economic challenges.* OECD Publishing.

OECD. (2021). *The inequality/environment nexus: Towards a people-centred green transition.* OECD Publishing.

Polanyi, K. (2001 [1944]). *The Great Transformation. The political and economic origins of our time.* Beacon Press.

Sanne, C. (2002). Willing consumers—Or locked-in? Policies for a sustainable consumption. *Ecological Economics, 42*(1–2), 273–287. https://doi.org/10.1016/S0921-8009(02)00086-1

Wade, R. H. (1996). Japan, the World Bank, and the art of paradigm maintenance: The East Asian Miracle in political perspective. *New Left Review I*, 2017.

World Bank. (2010). *World development report 2010. Development and climate change.* The World Bank.

World Bank. (2012). *Inclusive green growth. The pathway to sustainable development.* The World Bank.

World Bank. (2017). *Environmental and social framework.* The World Bank.

World Bank. (2021). *The economic case for nature. A global earth-economy model to assess development pathways.* The World Bank.

CHAPTER 7

Eco-Social Policy in Global Governance

This chapter theorises the empirical reconstructions and categorisations of both organisations' eco-social policy discourses (Chaps. 4 and 5). It puts forward four different but interconnected arguments. It argues, first, that the organisations' eco-social policy discourses are not coherent sets of ideas but instead propagate different sets of ideas shaped by what Polanyi (2001) has called the double movement (Sect. 2.3). This refers to the OECD's and World Bank's conceptions of green economy and green Keynesianism. While green economy resonates with the first movement towards establishing a self-regulating market and increasing the scope of private actors and market-based solutions, green Keynesianism is more attuned to the countermovement of calling for comprehensive government intervention and a profound commitment to social protection. This mechanism is discussed as a crucial one to stabilise and adapt the contemporary capitalist mode of production in the global governance system given ecological challenges. Second, the chapter assesses the OECD's and World Bank's eco-social policy discourses in terms of their potential to drive a social-ecological transformation. Both organisations still advocate substantial elements of "liberal globalism" (reflected in their green economy understandings) while as part of their understandings of green Keynesianism, they are moving towards a policy agenda of green embedded liberalism. They fall short of driving a social-ecological transformation that advances profound attempts at decommodifying nature and labour

R. Schulze Waltrup, *Eco-Social Policy in the OECD and the World Bank*, https://doi.org/10.1007/978-3-031-95429-0_7

and that invokes a post-growth perspective, including socio-economic democratisation, emphasising community (care) practices, selective economic deglobalisation and a cooperative global governance logic (Brand et al., 2020; Foundational Economy Collective, 2018; Novy, 2022). Third, the chapter argues that the OECD's and World Bank's eco-social policy discourse represents a "new critical orthodoxy" (Brand et al., 2020). This orthodoxy recognises that there are profound and far-reaching problems and crises, such as social inequality and the transgression of planetary boundaries, that would require some sort of large-scale societal and transformational change. However, it proposes solutions that are built on existing institutions and capitalist dynamics, which are assumed to be capable of addressing these crises (even though they contributed to creating them). In this process, some more radical perspectives are referred to but only to suggest an incremental adaptation of the given system. Green Keynesianism follows such a logic. It represents a move towards a "new" capitalist mode of production in which the state reclaims its control over the market in light of necessary policy changes in the context of ecological and social crises and how the market has contributed to them but reproduces the logic of capital accumulation and valorising nature. Fourth, in terms of policy paradigm change, the chapter argues that the struggle between green economy and green Keynesianism within the double movement dynamic might indicate a third-order paradigmatic change or at least a competition between third-order paradigms to become or stay dominant. Nevertheless, this falls short of a fourth-order change of "general systemic logics" (Buch-Hansen & Carstensen, 2021, p. 312). A fourth-order paradigm change would reject the economic-growth-based capitalist mode of production and move towards post-growth policy ambitions of radically decommodifying nature and labour. Due to the overarching struggles and fights among different policy perspectives, it cannot be argued that there is a coherent and unified "global eco-social policy paradigm". Instead, the eco-social policy ideas that both organisations promote are rather a "web of loose [eco-social] ideas" (Von Gliszczynski, 2015, p. 67ff.) in global governance that is shaped by the double movement dynamic.

7.1 The Double Movement in Eco-Social Policy

In order to explain the emergence of two opposing policy perspectives, green economy and green Keynesianism, in both organisations' policy discourses, this chapter draws on the concept of the double movement developed by Karl Polanyi in his book *The Great Transformation*. The first movement of the double movement consists of "the principle of economic liberalism" (Polanyi, 2001, p. 138), permeated by an economic laissez-faire approach that largely favours the interests of the trading class and is based on the propagation of free trade, while the second, also termed the countermovement, is driven by "the principle of social protection" (ibid.) that largely favours the interests of (rural) workers and policy interventions, such as protective legislation. Based on the notion of the double movement, Polanyi reconstructs the emergence and the establishment of the idea and practice of a self-regulating market as part of the first movement during the industrial revolution in England and highlights the detrimental social and ecological consequences that these specific market-operations created. As Polanyi argues, these detrimental consequences led to the emergence of a protective countermovement aiming to re-embed the market into society and prevent "the running of society as an adjunct to the market" (ibid., p. XXIV) with its destructive effects. The countermovement operates as "more than the usual defensive behaviour of a society faced with change" (ibid., p. 136). It is a protective movement that not only saves the social "fabric of society" (ibid., p. 37) but also, within a dialectic and dynamic interplay between both movements, stabilises the "organization of production that the market had called into being" (ibid., p. 136). It stabilises a capitalist mode of production, but one which is characterised by a comprehensive policy constellation of interventions and regulations. Such a countermovement led to the establishment of the embedded liberalism enshrined in the postwar order of comprehensive welfare state services and programmes in the Global North (Atzmüller et al., 2019; Deutschmann, 2019), which Heikki Patomäki (2014) sees as a synthesis of the dialectics of the double movement.

Contrary to Polanyi's expectations, the dynamics of the double movement did not cease with the institutionalisation of embedded liberalism. The experience of stagflation during the 1970s created profound scepticism concerning the ability of the existing institutions of embedded liberalism to effectively address and resolve pressing issues. It triggered a period of contestation and scrutiny of these institutions. Countercyclical fiscal policies and dependent central banks were then viewed as problematic. Concurrently, there was a rise in the influence of financial interests promoting neoclassical economic ideas that emphasised monetary stability and inflation control in macroeconomic policy-making as well as privatisation and deregulation. This shift contributed to a resurgence of a stronger self-regulating market order in the 1990s with striking similarities to the international economic system that had been identified as problematic in the 1920s and 1930s (Blyth, 2002, p. 5f.; see also Atzmüller et al., 2019; Deutschmann, 2019). Unsurprisingly, this resurgence led to profound social ruptures and a further exacerbation of activities resulting in an exceeding of planetary boundaries. Consequently, a countermovement (again) emerged that rejected the Washington Consensus's neoliberal policy prescriptions which advocated for deregulation, liberalisation, financialisation and (trade) liberalisation (Mendly, 2020; Patomäki, 2014).

Scholars such as Mendly (2020) and Patomäki (2014) have explicitly applied a Polanyian lens to analyse the global governance system. In fact, Dorottya Mendly (2020) argues that global governance has attained a hegemonic status precisely because it can accommodate the dynamics of the countermovement within its material structures and policy discourses. Thus, a critical balance exists within global governance between the establishment and perpetuation of market-oriented policies and a social protection approach that seeks to safeguard society from the adverse consequences of an unregulated market. Furthermore, global governance encompasses a diverse range of actors who embody different perspectives on the double movement (Mendly, 2020). However, she notes that, within the realm of global governance, market-based governance mechanisms tend to occupy a more dominant position, which means that interventions and regulations that aim to curb the operations of a self-regulating market are typically more reactive than anticipatory. Incumbent actors within global governance may incorporate concerns voiced by countermovements that advocate ecological preservation and mitigation of social pressure, but these efforts usually serve to reproduce, reinforce and adapt the existing capitalist mode of production (ibid.).

The notion of the double movement within global governance provides a framework for making sense of often complex and diverse policy discourses that seem to be mutually antagonistic at first glance. While Mendly (ibid.) proposes a rather general theoretical and historical analysis of the dynamic of the double movement in the global governance system, she admits that there is a need for more empirically grounded research on the subject. Quite briefly, and without getting into specific empirical detail, she notes with reference to the broad UN system that "[b]oth motivations of the double movement are represented, often 'under the same roof'" (ibid., p. 515). This book presents such a perspective empirically and reconstructs how both motivations of the double movement are represented under the same roof in the OECD and World Bank and their eco-social policy discourses.

More precisely, I argue that the green economy understandings of both IOs represent the first movement towards a (green) self-regulating market. However, there is a subtle difference. While the traditional conception of the first movement in Polanyi's framework entirely disregards ecological concerns, the green economy perspective recognises the importance of integrating ecological considerations into the operations of the market's first movement. Nevertheless, it closely aligns with market thinking, suggesting that significant government intervention to guide private actors and to steer the market is unnecessary and part of the problem. Instead, it calls for limited government intervention that specifically enables and empowers self-regulating practices among market participants. This aspect continues to reflect the core of the first movement's ideology.

It is axiomatic for green economy that the government's role should be restricted to providing a framework in which private actors and market-based instruments can flourish without taking on a strong intervening and steering role in driving economic developments. As laid out in Sect. 3.2.1, green economy follows the standard economics conception that this should not go beyond correcting market failures and externalities. The core is to price ecosystem services, internalising them into market costs and valuing them in economic and monetary terms (Tienhaara, 2018, p. 35). These are central themes within both organisations' eco-social policy discourses in which economic instruments are often framed as preferable to regulatory ones. The OECD, for example, values economic instruments for their flexibility in how to achieve a reduction of GHG emissions or resource throughput. It appreciates their market-friendliness in not harming competition and innovation (OECD, 2015, p. 17). Accordingly, it notes that "[t]axes encourage behavioural change but do not force change when it would be excessively costly" (ibid., p. 45). The World Bank, similarly, notes that carbon pricing should function "to incentivize commercial entities and citizens to reduce emissions and the private sector to co-invest in the key systems transitions, reducing the extent of additional public investment needed" (World Bank, 2021b, p. 11).

Green economy also argues that "[g]overnments are largely restricted to creating the conditions required for green markets to operate (e.g., protection of intellectual property rights, creation of permit trading schemes)" (Tienhaara, 2018, p. 37). In other words, governments are supposed to enact policies that serve private actors' interests and the market. This is reflected within the notion of "blended finance" that both organisations promote. The World Bank applies blended finance as "the strategic use of public finance for the mobilization of private capital flows towards emerging and frontier markets" (World Bank, 2021c, p. 112), while the OECD similarly observes a need for more "diversified risk mitigation tools, improved environmental risk analysis, and diversified financial instruments and models" (OECD, 2017, p. 37) that aim to mobilise private capital.

Polanyi (2001) argues that the first movement served the interests of the trading class and was based on the propagation of free trade. Green economy is also strongly connected with "greening the financial system" (see, in particular, the World Bank's green economy understanding as set out in Sect. 5.3.2). While this is to a certain extent an incorporation of ecological concerns into the interests of the trading class, it rejects large structural changes but focuses on inventing and promoting the establishment of new green markets while the basic market operations of capital accumulation that have caused these crises are reproduced and after all still appreciated. This takes place within a process of valorising nature while greening economic growth, so that traditional growth-dependent social policy can simply continue as before. However, the argument is that the market forces that are unleashed and empowered in the first movement (and this potentially continues with green economy) risk creating immense social and ecological pressures on society. This, as Polanyi argues, leads to an almost inevitable emergence of a countermovement based on government intervention and "the principle of social protection" (Polanyi, 2001, p. 138) that aims to re-embed the market into society, preventing "the running of society as an adjunct to the market" (ibid., p. XXIV). In eco-social terms, this countermovement is supposed to halt and prevent the commodification of nature and labour (Polanyi, 2001).

This book claims that both organisations' green Keynesian understandings represent such a countermovement that tries to even out green economy attitudes and re-embed the self-regulating market into society and nature more profoundly. The OECD and World Bank are not only organisations that promote the policy agenda of the first movement, they are also characterised by pressure, struggle and a dynamic that largely serves the interest of the countermovement. As argued in Chap. 3, green Keynesianism is a counter perspective to green economy in that it calls for comprehensive "interventions by the state comprising both transfer measures and public investments" (Custers, 2010, p. 174). Traditionally, neoclassical economics suggests that imperfections are due to the government's interventionist role. However, green Keynesianism provides a discursive shift in that this scapegoat role is allocated instead to "'dirty', corrupt, corporate capital and industrial polluters" (Goldstein & Tyfield, 2017, p. 10), while it approves of government taking a comprehensive role in public policy to the detriment of "dirty" capital and industries. In this shift, innovators and actors focusing on green and clean technologies are particularly empowered by the government. Green Keynesianism also resonates well with just transition approaches that highlight the role of the government in steering the transition towards a low-carbon economy and support disadvantaged groups with social protection measures (Krause et al., 2022).

Both organisations highlight the crucial role of the government in steering a just transition. As the World Bank notes, "The Bank will focus on protecting the poor in these reforms by strengthening social safety nets and facilitating communication campaigns to address political economy challenges" (World Bank, 2021b, p. 23). Such a perspective is also directly tied to "strong public policy" which "can help economies [to] absorb shocks of unavoidable climate impacts, minimise net social losses, and protect the welfare of those who most stand to lose" (World Bank, 2010, p. 60). The World Bank also appreciates green industrial policies which would aim "to green the productive structure of the economy by targeting specific industries or firms. They include industry-specific research and development subsidies, capital subsidies, and tax-breaks; feed-in tariffs; and import protection" (World Bank, 2012, p. 68). Furthermore, the Bank contextualises an increased appreciation of government intervention and steering markets within the development of the international system of the last decades:

> Over the past 30 years the role of the state has been cut back in various domains key to addressing the climate challenge, such as energy research. The retreat from direct intervention occurred with a switch from

> "government" to "governance" and an emphasis on the state's role in steering and enabling the private sector. This general trend hides a complex picture. Twentieth-century Europe saw various forms and degrees of state capitalism. The rise of East Asian economies, including China's, demonstrated the pre-eminence of the state in "governing the market" to deliver the most successful example of accelerated development. Most recently, the 2008 financial crisis showed the pitfalls of deregulation and unrestrained markets—and triggered renewed emphasis on bringing back the state. (World Bank, 2010, p. 330)

The OECD also clearly promotes crucial green Keynesian elements that speak to characteristics of the countermovement. The OECD also refers to just transition, which includes enabling "social dialogue, social capital investments, social protection, [and] skills and education investments" (OECD, 2021, p. 4), which is connected to "inclusive green growth" with a focus on social protection, "well-targeted income support measures, and upgrading skills to facilitate labour reallocation" (ibid.). One claim is that

> states must become more entrepreneurial, seeking to shape markets and steer the process of economic change, not simply correct market failures. An empowering and entrepreneurial state of this kind would allow the development of a new kind of social contract – a new relationship between the state, business, civil society and citizens. (OECD, 2020, p. 27f.)

Also, there is scepticism concerning the hegemonic economic frameworks that have persisted within the last few decades and mainly promote first movement policies:

> It would be surprising if it were just coincidence that so many deep problems have occurred at the same time. At the very least they surely demand that we seek to understand not just their individual causes but their interrelationships. And in doing so that we examine the nature of the economic system and economic policies from which they have arisen – or at least which have not prevented them. (ibid., p. 11)

The identified ambiguity in both organisations' approach to recomposing consumption and production highlights the dynamics of the double movement as well. Both organisations clearly focus on empowering consumers through non-intervening economic instruments, education, certification and labelling which primarily resonates with the first movement. But there is also some acknowledgment of communal modes of consumption and the role of government as a potential actor in driving market processes (countermovement). However, the latter is often framed to foster innovation, create leading markets and establish first-mover advantages, which again rather speaks to processes of commodification.

In sum, both organisations, in particular by promoting green economy and green Keynesianism partly within the same documents and partly within the same working units, inhabit positions characteristic of the first movement as well as of the countermovement. In other words, the establishment and development of eco-social policy in global governance are characterised by the double movement dynamic. In the following sections, the book expands on this with reference to the reproduction and reconfiguration of the existing dominant capitalist mode of production, suggesting that green Keynesianism, in a dialectical and dynamic interplay with green economy, stabilises the "organisation of production that the market had called into being" (Polanyi, 2001, p. 136). In order to develop this argument, the book also draws on the concept of a "new critical orthodoxy", which is also inspired by Polanyi (Brand et al., 2020).

7.2 Towards a Social-Ecological Transformation?

This section assesses the World Bank's and OECD's potential to drive a social-ecological transformation based on their eco-social policy ideas. Both still advocate substantial elements of "liberal globalism", which is largely reflected in their green economy understanding. However, as part of their green Keynesianism, they aim to reform their notion of liberal globalism, favouring an entrepreneurial state with comprehensive government intervention. Such reform attempts suggest a move towards green embedded liberalism, which is a rejection of both right-wing and left-wing attempts at deglobalisation and conceives the authority of public agencies as crucial to steering (transnational) capital and labour developments. Furthermore, in this section I contrast this vision of green embedded liberalism with more radical attempts to drive a social-ecological transformation that focuses on the decommodification of nature and labour (Brand et al., 2020; Novy, 2022), referring mainly to the proposals of the Foundational Economy Collective and other projects that follow social-ecological transformation criteria, such as socio-economic democratisation, emphasising community practices, selective de-globalisation and a cooperative global governance logic as also reflected in the post-growth perspective. I argue that the eco-social policy discourses of both IOs represent a new development phase within global governance's capitalist mode of production (reflected by green Keynesianism) but fall short of enacting a social-ecological transformation that puts the decommodification of labour and nature at its centre and moves towards post-growth-inspired policy-making.

Both IOs prescribe various policy ideas that reflect a policy agenda of "liberal globalism" (Novy, 2022). This is, in particular, the case with their green economy and recomposing consumption and production elements that subscribe to ideas of increasing the scope of private actors and appreciate market-based solutions as well as consumer sovereignty. These elements drive the market logic of liberal globalism. However, the green Keynesian understanding in particular of both organisations reflects an attempt at abandoning liberal globalism to a significant degree, and a move towards a policy-perspective of green embedded liberalism.

Liberal globalism promotes processes of globalisation and market-conforming democratisation. Within this thinking, appropriate global governance takes place within economic institutions and treaties, and "fierce economic competition and geopolitical rivalry" (Novy, 2022, p. 70) are dominant logics that structure practices of various actors within global governance. Liberal globalism aims to commodify and financialise substantial aspects of care practices, including health and education. The same holds true in relation to nature. Overall, there is the ambition to increase the scope of individuals, primarily as private agents and sovereign consumers. Liberal globalism has a cosmopolitan outlook and tries to fight any impediments to transnational exchanges, broadly favouring deregulation and liberalisation. Accordingly, government intervention is perceived as crucial solely in regard to ensuring personal and individual freedoms, while more comprehensive and profound interventions are perceived as inappropriate. Hence, according to liberal globalism, there should not be any value judgement by a community or the government on how individuals choose to live their personal life (Novy, 2022). This is also crucial part of both IOs' green economy understandings of favouring economic instruments based on which individuals can decide on their own whether and how they want to change their behaviour. The OECD, for example, appreciates the flexibility of these economic instruments. Similarly, the World Bank appreciates that they will "leave individual firms and households free to find the least-cost way to meet a climate target" (World Bank, 2010, p. 268).

Liberal globalism promotes a liberal economic order and involves the primacy of market-based solutions while rendering these solutions and ecological protection compatible with one another. Trade and technological progress are other important means to mitigate climate change while diffusing technologies worldwide. Or as Halliki Kreinin has put it critically: "The solution to the crisis is presented as the 'internalisation' of market 'externalities', the use of market mechanisms and the creation of new markets to solve crises caused by the market system itself" (Kreinin, 2020, p. 4). Such a perspective is also dominant in both organisations' understandings of green economy. As the World Bank states:

> Much of green growth is about good growth policies—addressing market failures and "getting the price right" by introducing environmental taxation, pricing environmental externalities (such as carbon pricing), creating tradable property rights, and reducing inappropriate subsidies. These measures are critical for enabling the private sector to undertake needed investments and innovations and for getting consumers to internalize the true costs of their behavior. (World Bank, 2012, p. 12)

There is also an explicit promotion of the creation of new markets, which not only extends to carbon markets (World Bank, 2020) but also acknowledges the economic value of biodiversity. Such markets would include "nature-based tourism supported by biodiversity, watershed services, and climate-regulating services. The main challenge [...] is to create markets for these services so that they become part of the visible economy and are efficiently provided" (World Bank, 2012, p. 117). One idea regarding the creation of markets for biodiversity practices and services is the provision of payment schemes that reward landowners if they actively engage in practices to protect the sink capacities of forests (World Bank, 2021a, p. XIII). The OECD also emphasises free trade as a crucial element in the transition to a low-carbon economy. This involves in particular the reduction of "[b]arriers to trade and investment [which] can place a serious break on the development and diffusion of green technologies globally" (OECD, 2011, p. 12). In sum, liberal globalism reflects the first movement within Polanyi's double movement dynamic. However, it is crucial to emphasise that the green liberal globalism promoted by the OECD and World Bank is nevertheless a shift from the uncompromising neoliberalism of previous epochs that at least to some degree takes account of ecological limitations. It is an integration of ecological considerations into basic liberal assumptions that Bernstein (2001) has also called "liberal environmentalism". Nevertheless, the issue with liberal globalism, as the countermovement suggests, is that its agenda of liberalisation and privatisation undermines the scope of public administrations, hindering them from providing comprehensive well-being for their constituents or citizens, and generally risks reproducing ecological and social crises. Given the strong neglect of social protection and collective modes of organising nature-society relations, liberal globalism, viewed from a Polanyian perspective, provokes the emergence of countermovements that can take various forms, either reactionary or progressive (Novy, 2022).

Considering this, I argue that the IOs' green Keynesian perspective reflects attempts to reform and counter the policy agenda of "liberal globalism" and, as discussed in the previous section, echoes countermovement dynamics. While there are certainly elements of liberal globalism within both IOs' green economy understandings, their green Keynesian understandings entail policy prescriptions to ensure a "just transition to ecological modernisation" (Kreinin, 2020), which centres around the government and its potential to steer a just transition, which rather goes against key policy principles of "liberal globalism": Here, governments abandon the mantra of non-intervention and instead drive such a transition by pushing certain technological developments and providing comprehensive community services and social protection. Therefore, there is increased focus on government control in regard to industrial action and innovation. Generally speaking, there is also more awareness of social protection in terms of the potential regressive effects of climate mitigation measures (Kreinin, 2020). Such a policy approach rejects core assumptions of liberal globalism. The two perspectives of green economy and green Keynesianism prescribe contrary roles for the government within an economic framework. Hence, both organisations do not simply follow the policy agenda of liberal globalism but instead are busy reforming it and transitioning to a policy agenda of green embedded liberalism.[1] Green embedded liberalism is a rejection of right-wing and left-wing deglobalisation movements. While there is agreement to resist current forms of (economic) globalisation, this does not necessarily need to come with a

[1] To describe crucial elements of green embedded liberalism, this section refers to the description of what Bishop and Payne (2021) have called "re-embedded post-neoliberal reglobalisation".

complete rejection of the process, but instead with reformative ambitions. Green embedded liberalism recognises that the neoliberal-dominated type of globalisation is a historically specific one and that there can be different forms. It calls for stronger government intervention, and a redefinition of the narrative that free trade and liberalisation would be a panacea for the pressing challenges of modern society. Instead, it highlights that there is a need for a redistribution of resources within and between countries that might appear due to liberalisation. Its referring to the idea of "embedded liberalism" (Ruggie, 1982), which suggests a system in which highly embedded "public agencies have the authority to intervene in capital, labour products and other markets" (Fioretos & Heldt, 2019, p. 1097), is in contrast to the disembedded systems that have been promoted within the neoliberal process of globalisation and liberal globalism, in which "public agencies lack effective means to structure the behaviour of market participants" (ibid.). This way of organising policy-making can take different forms. On a global level, it can come with policy scripts that legitimise the government "to pursue legitimate social [and ecological] purposes and enjoy the necessary national policy space to manage their economic development successfully, without fear of retreat into a regressive nationalism [...] to ensure the fairest distribution of gains over time" (Bishop & Payne, 2021, p. 5). This shift away from a policy agenda of liberal globalism is reflected in both organisations' green Keynesian understandings (Sects. 4.3.3 and Sect. 5.3.4) and, quite remarkably, in both of the following statements:

> as the prime repositories of political legitimacy, governments will be expected to steer the private sector, facilitate community action, and establish the optimal decentralization of adaptation and mitigation decision making and action. On top of steering, governments will be expected to play an "ensuring" function: guaranteeing that targets and goals are achieved through new emphasis on regulation, taxation, long-term planning, and communication. (World Bank, 2010, p. 331)

Or, as the OECD puts it:

> The neoclassical framework presupposes that well-functioning markets optimise overall welfare, and government policy is therefore justified to correct market failures. But if public policy is to aim at different ethical outcomes, the state will have to play a larger role in guiding, or steering, the overall patterns of economic activity to achieve them. Through public service and

> welfare provision it can also support a fairer and more productive form of economic development. "Correcting market failure" will not be sufficient; markets can also be "shaped" in pursuit of publicly-determined goals. (OECD, 2020, p. 23)

This suggests a move towards embedded systems like those extant during the era of "embedded liberalism", while green Keynesianism indicates an explicit green conditionality to these embedded systems.[2] However, green Keynesianism still relies on economic growth and generally on conceptions of full employment that are connected with mass-consumption and production (though of green and sustainable products), which is deemed problematic *tout court* from a post-growth perspective. From this (and also from a Polanyian) perspective, abandoning such a capitalist mode of production would come down to a profound rejection of the commodification of nature and labour (that green Keynesianism still promotes to a certain degree; see the following section). This would be crucial, as such a commodification of nature and labour has helped reactionary fascist regimes in Europe to gain power (Polanyi, 2001).[3] The commodification of nature, for instance, suggests that there is more or less unlimited access to natural resources as long as someone can pay for it (Brand et al., 2020, p. 169). The OECD and World Bank do not question the commodification of nature but instead reproduce and promote it, as is most importantly

[2] Countermovements in response to the institutionalisation of disembedded systems can take different forms, from reactionary to progressive (Polanyi, 2001). Generally, the OECD and World Bank reject a nationalistic and reactionary policy agenda, which would be, for example, reflected in a one of "nationalistic capitalism" (Novy, 2022). Such a perspective would focus strongly on capitalist values, such as property rights, competition and innovation. Nevertheless, contrary to liberal globalism (which also cherishes these values), nationalistic capitalism would aim to withdraw from key liberal global governance organisations and entail an imperialistic and nationalistic policy agenda shaped by racist thinking and cleavage thinking between the Global North and South. It would entail challenging Darwinism with religious fundamentalism, or questioning scientific advances, such as knowledge of climate change or vaccines. Generally, it follows the idea that enjoyment of the good life can only be attained by a selected few (usually based on national accounts). This is contrary to liberal globalism that promotes a good life for all based on fulfilment of human rights. Nationalistic capitalism rejects forms of globalisation that advocate multiculturalism, migration and "unfair" economic competition. Under fair conditions of competition, it suggests that nationalistic economic and political superiority would manifest themselves: "It combines dominium, domination via property rights, and imperium, domination via political force" (Novy, 2022, p. 72).

[3] Note the similarity with our present time.

reflected in both organisations' green economy understandings and their focus on establishing market-based solutions, such as cap-and-trade schemes.

Decommodifying nature means promoting a different way of organising the relationship between nature and society. Here, I refer to proposals from the Foundational Economy and other political projects that represent a transformative countermovement that might be able to find an alternative way of organising the society–nature relationship and, thus, bring about a provisioning system that accomplishes social need satisfaction while staying within planetary boundaries through a process of decommodifying nature and labour (Foundational Economy Collective, 2018; Novy, 2022). Such a social-ecological transformation would be based on planetary coexistence, socio-economic democratisation, selective economic deglobalisation and a cooperative governance logic.

A crucial element of a social-ecological transformation is socio-economic democratisation, which is supposed to come along with a prioritisation of need satisfaction based on "less tradeable commodities" (Novy, 2022, p. 73). Accordingly, there need to be context-aware policies that deal with certain sectors, such as the commodity sector and the caring practice sector, in different ways. The way electronic devices are consumed should be regulated differently from how basic and important need-satisfiers (in terms of goods and services) are consumed. The foundational economy implies a strong prioritisation of equality and sustainable modes of living usually based on community and collective practices. Here, collective provisioning is crucial for the satisfaction of basic needs as it is primarily collective undertakings that can guarantee that everyone receives the same amount of need satisfiers and that no individuals consume to the detriment of other individuals. This means that the policy development of the financialisation and commodification of basic care practices needs to be reversed. There needs to be a shift away from private consumption in order to favour "common, public and regionalised provisioning of basic goods" (Novy, 2022, p. 73). It promotes a change of basic modes of living, in particular by promoting decommodification and resource sharing and collective provisioning of goods and services, which will, in the long run, support equality. There need to be clear commitments to focus on key provisioning systems, such as health and education, as well as compensating losses. This also entails measures "to combat wealth concentration and luxury consumption" (Novy, 2022, p. 75). Certain aspects of deglobalisation would be crucial, including containing the pressure to liberalise provisioning systems, expand high-carbon infrastructure ((air)ports) while at the same time still “strengthening cooperation-oriented forms of globalisation” (Novy, 2022, p. 75). Infrastructural configurations within a foundational economy differ depending on the social and geographical context, but, generally, they put the logic of decommodification centre stage. They speak to the delivery of services and practices that are essential for everyone’s well-being: from electricity, water and housing to local food supply, health and care (Bärnthaler et al., 2023, p. 118). They would be “accessible (e.g. public spaces), affordable (e.g. housing, health, and care) and sustainable, providing inter alia green spaces, decarbonized housing and modes of mobility as well as sustainable energy” (ibid.).

Similarly, Brand & Wissen (2018, p. 80f.) highlight different aspects that are important for fertilising a social-ecological transformation and overcoming the “imperial mode of living” (further discussed in the following section). First, everyday experiences need to be changed and shaped by a different social logic. For example, the value of leisure could be put centre stage, which would come with a reduced role for wage labour in an individual’s identity. Second, there would be a need to strengthen processes of “deceleration and use value-oriented production” (ibid.) and democratisation. Democratisation can also be accelerated by local collectives and a decentralised organisation of energy supply and, more generally, decision-making, encouraging local consumption patterns. Third, it would be necessary to couple production and reproduction. Workers would need to be integrated into the process of producing high-value goods that are crucial for their reproduction. This must come with a prevention of the overconsumption and overproduction that has at its centre the maximisation of profits:

> the structural imperative to perpetually create new needs as soon as the means to satisfy them have been developed has to be overcome; efforts should be directed to democratically negotiate what and how much is necessary for a good live for all in line with the reproductive requirements of nature and constantly posing the question of what might *not* be needed. (Brand & Wissen, 2018, p. 103)

These transformative requirements are also well captured in the post-growth perspective (Sect. 3.4) that clearly abandons the notion of increasing profits and centres around caring for each other: “not profits, productivity and power but the caring for oneself, for others, for societies and nature should be the central reference of thinking and action" (Brand & Wissen, 2018, p. 104). This implies, for example, work time reduction policies, job guarantees, local currencies, egalitarian governance structures for organisations based on "one-member-one-vote", communal property rights regimes, minimum income floor and maximum ceilings and wealth caps (Sect. 3.4.3).

Needless to say, this book's focus is on IOs that provide specific policy recommendations to national governments, which obviously entails a concentration on the (limited or advanced) role of the government in policy-making. The processes that led to the great transformation described by Polanyi were quite diverse and not completely steered by governments. Nevertheless, the state still implemented various small-scale changes that accelerated the transformation, suggesting that government does indeed play a crucial role in driving or preventing transformational attempts. Governments have the capacity to "orchestrat[e] and/or enabl[e] multiple socio-technical transitions" (Eckersley, 2021, p. 248). Furthermore, the nation-state is a crucial institution that can regulate material flows and re-embed a self-regulating market into social-ecological community constellations. There is no institution in existing polities that has the potential to redistribute financial flows to a similar extent. Therefore, the nation-state might have a crucial role in mitigating apparent inequalities within transition processes (ibid.).

Nevertheless, alternative visions of a sustainable mode of living can only be institutionalised through societal struggles and new experiences and practices on various scales including different actors that create different conceptions of how society can be organised in relation to nature. Large-scale changes might only be possible as a result of "radically different daily practices of individuals and collectives" (Brand & Wissen, 2018, p. 76). This suggests opposition to high-carbon infrastructure and widespread rejection of unsustainable individual practices. Hence, these processes can hardly be steered by the government alone. Instead, we also need to acknowledge that governments are often characterised by certain vested interests that impede such transformation processes (Brand & Wissen, 2018). A social-ecological transformation can only be carried out based on a large-scale coalition of different societal actors with a strong focus on local communities and cooperatives that potentially transcend the nation-state.

7.3 Eco-Social Policy as a New Critical Orthodoxy

The OECD's and World Bank's eco-social policy prescriptions bear striking similarities with what Brand et al. (2020) have conceptualised as a "new critical orthodoxy". This orthodoxy recognises that there are profound and far-reaching problems, such as social inequality and ecological collapse, that require some sort of large-scale transformative societal change but then scales down the proposed change to a fairly gradual and incremental adaptation of existing institutions and practices.

The need for large-scale or "transformational" change is often depicted in the introductions to the IOs' documents that detail eco-social policy concepts. The OECD report *The Inequalities–Environment Nexus. Towards a People-Centred Green Transition* states in its introduction:

> Most of the environmental challenges are global in terms of drivers, impacts and prevalence. They are also long-term and structural, and pose a serious risk of irreversible damage to the natural capital base on which future prosperity and well-being of all generations depend. Action is needed to catalyse a systemic and deep transformation in the ways we produce, consume and behave in order to "build back better", i.e. not only getting economies and livelihoods back on their feet, but also safeguarding prosperity for the longer term. (OECD, 2021, p. 9)

The World Bank, in its WDR on climate change, states:

> Tackling the immense and multidimensional challenge of climate change demands extraordinary ingenuity and cooperation. A "climate-smart" world is possible in our time–yet, as this report argues, effecting such a transformation requires us to act now, act together, and act differently. (World Bank, 2010, p. XIII)

However, it is characteristic of the "new critical orthodoxy" to then go on to propose solutions that rely on existing institutions and capitalistic dynamics that are assumed to be capable of solving contemporary crises and problems if they are adapted in specific ways (Brand et al., 2020, p. 163f.), which, as argued below, is most importantly reflected by the move towards green Keynesianism by both organisations.

Accordingly, the eco-social orthodoxy of the OECD and World Bank does not refer to more radical approaches that explicitly question incumbent institutions that have created the contemporary mode of production and living, such as the nation-state, the market or prevailing economic logics, such as economic growth and competition. Moreover, the inherent trait of capitalism to look for new profitable investments and further commodify nature is not addressed as being problematic. Instead, the orthodoxy appreciates this trait as a key solution to pressing problems. Resource extractivism prevails to a large extent, which is important in producing new and competitive technologies. Furthermore, the orthodoxy suggests that there should be a steering government that would be better equipped than the market to deal with certain problems. However, there is no critical discussion of how nation-states are shaped by certain power constellations that might hinder dealing effectively with ecological and social problems (Brand et al., 2020, p. 163f). Instead, a steering government is presented as a key solution to those problems.

Nevertheless, the current policy discourses, in particular green Keynesianism, that both organisations advocate, represent a new development phase within the global governance's capitalist mode of production and a change of ideas about how a decent economy has to function and what kind of role the government has, namely a steering role based on ecological conditionality instead of a role that provides a framework for privatisation, deregulation and liberalisation. Within this process, there are profound and complex negotiations between different ideas of how an economy should operate. More radical views that want to overcome capitalism also emerge and have a certain relevance within this struggle. Cases in point are the ambitions of, for example, NAEC or WISE within the OECD to go "beyond GDP" and to limit the meaning of economic growth as one key principle of (economic) policy-making. But incumbent forces often take certain elements of these more radical perspectives and integrate them into existing structures. Here, basic characteristics of capitalism may be preserved, although some changes take place in terms of the institutional conditions governing how capitalism operates (Brand et al., 2020, p. 166), which in the OECD's discourse is reflected, for example in the government's role in achieving different kinds of objectives that go beyond simply creating economic growth:

> For it is in the patterns of investment and forms of production themselves that the major problems and challenges arise. If we are to achieve the new economic goals we have set out – environmental sustainability, improved wellbeing, a reduction in inequality, and greater resilience – these need to be

> built into the structures of the economy from the outset, not simply hoped for as a by-product, or added after the event. (OECD, 2020, p. 25)

This "new" capitalist mode of production does not problematise the "imperial mode of living" (Brand & Wissen, 2018), which describes the normalisation of unequal exchange relations between the Global North and South that rest on an organisation of mass production and consumption as well as a relatively stable set of welfare institutions in the Global North that rely in turn on ecological degradation to uphold the latter's dominant position. Neither does it problematise the Western way of production and consumption and how it became universalised in processes of globalisation. A Western model of this kind requires an "imperial mode of living" that normalises access to cheap products in everyday life, which requires resource extractivism and cheap labour, often from the Global South (Brand & Wissen, 2018; Hickel et al., 2022). It also entails conceiving Western everyday life as desirable in the process of urging developing countries to become "developed" or "modern". Accordingly, there is rarely a problematisation of a Western lifestyle being adopted by the upper and middle classes of developing countries (Brand et al., 2020, p. 164). This is also reflected in the OECD's and World Bank's attempts to promote "first-mover advantages". Accordingly, other countries could simply follow these "new practices" and technologies that are promoted globally or, to use another word, "universalised" but with strong protection for private property rights. This means that the "imperial mode of living" and the green growth narrative incentivise developing countries to develop into the model represented by the Global North. Within this process of global development and the aspiration towards it, regions become dependent on raw materials that are necessary to produce competitive and innovative (green) technology. There are obviously limits to resource availability: "societies seem to be approaching the *limits to capitalist nature*" (Brand & Wissen, 2018, p. 3), and there are different configurations presented of how these limits can be managed.

In that regard, green Keynesianism is strongly tied to a policy agenda of discursively expanding planetary boundaries in time and space. Green Keynesianism represents a form of capitalism that does not rely on fossil fuels as its main resource input but instead on biomass and metals. The assumption is that a green capitalism of this kind "can cope with biophysical scarcities and environmental destruction by discovering and valorising new resources, substituting old ones and opening up new territorial and social spaces for capital accumulation" (Brand & Wissen, 2018, p. 4). This is quite markedly reflected within elements of the OECD's green growth understanding that speak to creating a "knowledge economy" while emphasising innovation as a crucial means to move away from the frontier at which the use of natural capital impedes growth patterns: "By pushing the frontier outward, innovation can help to decouple growth from natural capital depletion" (OECD, 2011, p. 10). In other words, the pressure to invoke scarcity of resources in particular with regard to GHG emissions does not automatically lead to a decline of capitalism. Quite the contrary, ambitions to reform capitalism are characterised by practices of valorising new elements of nature. Accordingly, green Keynesianism is on the hunt for new accumulation strategies that resemble crisis management, which results in a top-down nation-state-steered change of capitalism. Clearly, this represents an alternative pathway to the previous one favouring deregulation and privatisation. Nevertheless, the "new" steering, entrepreneurial nation-state also relies on economic growth to increase revenues based on taxes. Furthermore, the BRICS (Brazil, Russia, India, China, South Africa) countries, for example, are eager to increase their resource use in order to "grow" and receive increased geopolitical relevance, expanding the overall need for resources on a global level (Brand & Wissen, 2018). These developments are symptomatic of capitalist crisis management that reproduces and strengthens the continuing quest for a "valorisation of nature, and an economisation of ecological crisis management which both may open up new – albeit socially and spatially exclusive – growth potentials" (Brand & Wissen, 2018, p. 47).

The application of a "new critical orthodoxy" as part of the OECD's and World Bank's eco-social policy discourses suggests that there is a lack of profound criticism that could be targeted at the systemic operations of the global political economy. In fact, this is a crucial mechanism employed by incumbent actors that reproduces and adapts the capitalist mode of production in global governance to stay dominant while at the same time obscuring the emergence and potential of more radical political pathways that would profoundly challenge the incumbent economic order. This also emphasises how attempts to integrate environmental and social policy in an eco-social policy take place within the operation of the global political economy that pressures everyone to construct the ecological and social sphere to be in service to the economy, most importantly, the existing capitalist mode of production and economic growth (see also Schulze Waltrup et al., 2023).

7.4 Towards a Global Eco-Social Policy Paradigm?

This final section assesses the empirical reconstructions of the OECD's and World Bank's eco-social policy discourses in terms of the extent to which a paradigm change may be taking place as part of eco-social policy discussions. The overarching goal of established policy-making constitutes a policy paradigm to structure the way that policy-makers perceive policy problems and their solutions, how they communicate about them and which words they should use: "Like a *Gestalt*, this framework is embedded in the very terminology through which policymakers communicate about their work, and it is influential precisely because so much of it is taken for granted and unalienable to scrutiny as a whole" (Hall, 1993, p. 279). Different transition and transformation discourses entail different ideas on what is seen to be a problem within the economy and therefore also entail different conclusions as to what needs to be changed. In this sense, these discourses are part of different policy paradigms, which means that they entail different policy objectives concerning different policy problems that policy-makers are supposed to address (Buch-Hansen & Carstensen, 2021, p. 309). Facts and evidence that would potentially undermine the policy paradigm are often simply overlooked, ignored or reinterpreted to make them congruent with the existing paradigm (Princen & Van Esch, 2015, p. 357f.)

In Peter Hall's original 1993 work on policy paradigms, the broadest and most consequential level of change is third-order change. However, this is still conceptualised as change within a capitalist framework—from Keynesianism to monetarism, both driven by economic growth. Alternatively, in Polanyi's words (Sect. 2.3), this signals a change from the dominance of a countermovement (Keynesianism reflecting the counter-movement that established embedded liberalism) to the dominance of the first movement (monetarism reflecting the idea of establishing a self-regulating market without much government intervention). Noting shortcomings in terms of conceptualising potential policy change in this original version, Buch-Hansen and Carstensen introduced a fourth level of general systemic logic: "A logic is systemic if it is permanent and general: while institutions forming part of a given economic system may change, the underlying logic does not (as long as that system is in operation)" (Buch-Hansen & Carstensen, 2021, p. 312). Accordingly, third-order change does not change the logic of the existing capitalist system, instead it changes the ideas of how a decent and appropriate capitalist system functions—for example, in regard to the role of the government in economic developments. A fourth-order change, on the other hand, brings in a different system that abandons key principles of capitalism and represents "much deeper institutional and ideational change" (Buch-Hansen & Carstensen, 2021, p. 312). In that regard, the double movement dynamic between green economy and green Keynesianism (Sect. 7.1) might introduce a third-order paradigm change or, at the very least, represent a competition between third-order paradigms (green economy and green Keynesianism) vying for institutionalisation. This is mainly reflected within an appreciation and advocacy of green Keynesianism and green embedded liberalism that are opposed to ideas of green economy and liberal globalism that only accept government intervention to the degree that it enables deregulation and privatisation. However, such a change would still fall short of a fourth-order one that would entail a change of "general systemic logics" (Buch-Hansen & Carstensen, 2021, p. 312). Fourth-order change would abandon the contemporary capitalist mode of production and, for example, embark on post-growth policy-making that explicitly rejects a valued notion of economic growth and the capitalist system. Such fourth-order change would come, for example, with a transition towards post-growth policy ambitions of radically decommodifying nature and labour (Sect. 3.4) and different criteria for a social-ecological transformation, namely socio-economic democratisation emphasising community practices, selective economic de-globalisation and cooperative global governance logic (Sect. 7.2).

The simultaneous existence of the green economy and green Keynesianism, or the dynamic interplay of the double movement in both organisations' policy discourses, indicates that paradigm change is not necessarily, as Hall (1993) suggests, an abrupt process but instead takes place quite gradually (Carson et al., 2009; Princen & Van Esch, 2015; Skogstad & Schmidt, 2017). This also rejects Hall's original conception of paradigms as having hardly any chance of coexisting next to each other since they would entail incommensurable assumptions that could not be remedied. Other research also suggests that it is likely that there are mutually opposing paradigms simultaneously existing within a given policy context while one of them can still be the dominant one (Princen & Van Esch, 2015). Sabatier and Jenkins-Smith (1993) similarly suggest that dominant paradigms (in their word belief systems) compete with minority belief systems. The latter constantly try to challenge and contest the dominant system by referring to various cognitive concepts and normative ideas

(Skogstad & Schmidt, 2017, p. 9). NAEC within the OECD also explicitly invokes the notion of a paradigm shift to push its policy agenda:

> Social scientists describe these moments of economic change as "paradigm shifts" – periods when old orthodoxies are unable either to explain or to provide policy solutions to conditions of crisis, and new approaches take their place. More than a decade after the financial crisis, with the global economy and many individual OECD countries facing multiple crises, our argument is that the time is ripe for another such paradigm shift. The frameworks and prescriptions which have dominated policy making in recent decades are no longer able to generate solutions to the problems and challenges we face today. We need a less incremental, more profound form of change. (OECD, 2020, p. 29)

Rhetorically, claiming to pursue a paradigm shift while at the same time suggesting green Keynesianism-inspired policy objectives that aim at reproducing and adapting the capitalist mode of production curtails other more radical policy developments and possible paradigm shifts that would signal a fourth-order paradigm change of "general systemic logics" (Buch-Hansen & Carstensen, 2021, p. 312), such as more transformative post-growth policies (see also Schulze Waltrup, 2024). Nevertheless, within the OECD, NAEC is already pursuing a relatively progressive policy agenda that reflects the policy principles of the countermovement. Other units, such as the Economics Department, reflect policy principles of the first movement that discuss GDP-projections and monetary policy with tight public budgets as crucial guidance for policy-makers (OECD, 2023). This suggests that there is a struggle between paradigms or movements within the OECD between different units also in terms of relevant eco-social policy notions (not only green economy and green Keynesianism). There is no coherent and unified "global eco-social policy paradigm". Instead, the eco-social policy constellations of ideas that both organisations promote are a "web of loose [eco-social] ideas" (Von Gliszczynski, 2015, p. 67ff.) in global governance that is part of, and shaped by, the double movement dynamic, meaning that it is also not internally coherent. Accordingly, such a web is created out of various ideas and connections between the two movements. It also takes account of slightly different meanings of policy agendas between actors: The OECD has a more profound green Keynesian understanding with different conceptions and emphases than the World Bank, while the World Bank has a more sophisticated mandate concerning green economy (Chap. 6).

References

Atzmüller, R., Aulenbacher, B., Brand, U., Décieux, F., Fischer, K., & Sauer, B. (Eds.). (2019). *Capitalism in transformation. Movements and countermovements in the 21st century*. Edward Elgar.

Bärnthaler, R., Novy, A., & Stadelmann, B. (2023). A Polanyi-inspired perspective on social-ecological transformations of cities. *Journal of Urban Affairs, 45*(2), 117–141. https://doi.org/10.1080/07352166.2020.1834404

Bernstein, S. (2001). *The compromise of liberal environmentalism*. Columbia University Press.

Bishop, M. L., & Payne, A. (2021). The political economies of different globalizations: Theorizing reglobalization. *Globalizations, 18*(1), 1–21. https://doi.org/10.1080/14747731.2020.1779963

Blyth, M. M. (2002). *Great transformations. Economic ideas and institutional change in the twentieth century*. Cambridge University Press.

Brand, U., & Wissen, M. (2018). *The limits to capitalist nature. Theorizing and overcoming the imperial mode of living*. Rowman & Littlefield.

Brand, U., Görg, C., & Wissen, M. (2020). Overcoming neoliberal globalization: Social-ecological transformation from a Polanyian perspective and beyond. *Globalizations, 17*(1), 161–176. https://doi.org/10.1080/14747731.2019.1644708

Buch-Hansen, H., & Carstensen, M. B. (2021). Paradigms and the political economy of ecopolitical projects: Green growth and degrowth compared. *Competition & Change, 25*(3–4), 308–327. https://doi.org/10.1177/1024529420987528

Carson, M., Burns, T. R., & Calvo, D. (2009). Introduction. In M. Carson, T. R. Burns, & D. Calvo (Eds.), *Paradigms in public policy. Theory and practice of paradigm shifts in the EU* (pp. 11–28). Internationaler Verlag der Wissenschaften.

Custers, P. (2010). The tasks of Keynesianism today: Green new deals as transition towards a zero growth economy? *New Political Science, 32*(2), 173–191. https://doi.org/10.1080/07393141003722032

Deutschmann, C. (2019). Karl Polanyi as a theorist of disembedded markets. In R. Atzmüller, B. Aulenbacher, U. Brand, F. Décieux, K. Fischer, & B. Sauer (Eds.), *Capitalism in transformation. Movements and countermovements in the 21st century* (pp. 60–74). Edward Elgar.

Eckersley, R. (2021). Greening states and societies: From transitions to great transformations. *Environmental Politics, 30*(1–2), 245–265. https://doi.org/10.1080/09644016.2020.1810890

Fioretos, O., & Heldt, E. C. (2019). Legacies and innovations in global economic governance since Bretton woods. *Review of International Political Economy, 26*(6), 1089–1111. https://doi.org/10.1080/09692290.2019.1635513

Foundational Economy Collective. (2018). *Foundational economy. The infrastructure of everyday life*. Manchester University Press.

Goldstein, J., & Tyfield, D. (2017). Green Keynesianism: Bringing the entrepreneurial state back in(to question)? *Science as Culture, 27*(1), 74–97. https://doi.org/10.1080/09505431.2017.1346598

Hall, P. A. (1993). Policy paradigms, social learning, and the state: The case of economic policymaking in Britain. *Comparative Politics, 25*(3), 275–296. https://doi.org/10.2307/422246

Hickel, J., Dorninger, C., Wieland, H., & Suwandi, I. (2022). Imperialist appropriation in the world economy: Drain from the global south through unequal exchange, 1990–2015. *Global Environmental Change, 73*, 102467. https://doi.org/10.1016/j.gloenvcha.2022.102467

Krause, D., Stevis, D., Hujo, K., & Morena, E. (2022). Just transitions for a new eco-social contract: Analysing the relations between welfare regimes and transition pathways. *Transfer, 28*(3), 367–382. https://doi.org/10.1177/10242589221127838

Kreinin, H. (2020). *Typologies of 'just transitions': Towards social-ecological transformation*. Ecological Economic Papers, 35., WU Vienna University of Economics and Business, Vienna.

Mendly, D. (2020). Global governance and the double movement. *Global Governance: A Review of Multilateralism and International Organizations, 26*(3), 500–521. https://doi.org/10.1163/19426720-02603006

Novy, A. (2022). The political trilemma of contemporary social-ecological transformation—Lessons from Karl Polanyi's *The Great Transformation*. *Globalizations, 19*(1), 59–80. https://doi.org/10.1080/14747731.2020.1850073

OECD. (2011). *Towards green growth*. OECD Publishing.

OECD. (2015). *Creating incentives for greener products: A policy manual for eastern partnership countries*. OECD Publishing.

OECD. (2017). *Investing in climate, investing in growth*. OECD Publishing.

OECD. (2020). *Beyond growth: New approaches to economic challenges*. OECD Publishing.

OECD. (2021). *The inequality/environment nexus: Towards a people-centred green transition*. OECD Publishing.

OECD. (2023). *OECD economic outlook, interim report March 2023: A fragile recovery*. OECD Publishing.

Patomäki, H. (2014). On the dialectics of global governance in the twenty-first century: A Polanyian double movement? *Globalizations, 11*(5), 733–750. https://doi.org/10.1080/14747731.2014.981079

Polanyi, K. (2001 [1944]). *The Great Transformation. The political and economic origins of our time*. Beacon Press.

Princen, S., & Van Esch, F. (2015). Paradigm formation and paradigm change in the EU's stability and growth pact. *European Political Science Review, 8*(3), 355–375. https://doi.org/10.1017/s1755773915000089

Ruggie, J. G. (1982). International regimes, transactions, and change: Embedded liberalism in the postwar economic order. *International Organization, 36*(2), 379–415. https://doi.org/10.1017/s0020818300018993

Sabatier, P. A., & Jenkins-Smith, H. C. (Eds.). (1993). *Policy change and learning: An advocacy coalition approach*. Westview Press.

Schulze Waltrup, R. (2024). The OECD, global challenges and contestation of the economic growth paradigm. In A. Isakova, M. Neuwinger, R. Schulze Waltrup, & O. Uraiqat (Eds.), *Constructing global challenges in world politics* (pp. 214–227). Routledge.

Schulze Waltrup, R., Moore, M., & Paulsen, T. (2023). Eco-social policy in the global political economy: Analysing shifting discourses on agricultural subsidies. *European Journal of Social Security, 25*(4), 506–522. https://doi.org/10.1177/13882627231206002

Skogstad, G., & Schmidt, V. A. (2017). Introduction: Policy paradigms, transnationalism, and domestic politics. In G. Skogstad & V. A. Schmidt (Eds.), *Policy paradigms, transnationalism, and domestic politics* (pp. 3–35). University of Toronto Press.

Tienhaara, K. (2018). *Green Keynesianism and the global financial crisis*. Routledge.

Von Gliszczynski, M. (2015). *Cash transfers and basic social protection*. Palgrave Macmillan.

World Bank. (2010). *World development report 2010. Development and climate change*. The World Bank.

World Bank. (2012). *Inclusive green growth. The pathway to sustainable development*. The World Bank.

World Bank. (2020). *Transformative climate finance: A new approach for climate finance to achieve low-carbon resilient development in developing countries*. The World Bank.

World Bank. (2021a). *The economic case for nature. A global earth-economy model to assess development pathways*. The World Bank.

World Bank. (2021b). *Climate change action plan 2021–2025. Supporting green, resilient and inclusive development*. The World Bank.

World Bank. (2021c). *Toolkits for policymakers to green the financial system*. The World Bank.

CHAPTER 8

Conclusion

This book aims to provide the groundwork for a new research field in global eco-social policy. Most attempts to discuss global eco-social policy have been "awareness-raising" in the sense that they highlight environmental policy considerations and how they (potentially) interact with global social policy arrangements. Even though global social policy researchers have identified environmental sustainability as one of the biggest issues for the field (Smyth & Deeming, 2018, p. 325), there is hardly any empirical research on the integration of environmental and social policy at the global level of IOs (but see Cigna et al., 2023; Schulze Waltrup et al., 2023). The few empirical attempts have produced general observations, examining how IOs that traditionally have a narrow social policy focus have started to incorporate climate change issues in their respective social policy agendas (Lakeman, 2021) or discussed the Sustainable Development Goals as an exemplary eco-social policy, criticising their moderate transformative potential (Köhler, 2016). This book is the first to provide an in-depth eco-social policy assessment of crucial actors within global governance, namely the OECD and the World Bank, from an eco-social policy perspective. Accordingly, it provides various interconnected contributions to different literature strands, most importantly eco-social policy, global social policy and global governance, as this conclusion outlines below.

R. Schulze Waltrup, *Eco-Social Policy in the OECD and the World Bank*, https://doi.org/10.1007/978-3-031-95429-0_8

While existing eco-social policy literature lacks conceptual clarity as to what an eco-social policy perspective is, and how such a perspective should be applied in empirical research, this book has developed an eco-social policy typology for that purpose (see also Schulze Waltrup, 2025). Before the typology proposed and discussed different eco-social policy perspectives, including the green-growth-inspired, but conflicting perspectives of green economy and green Keynesianism, as well as recomposing consumption and production, and post-growth, this book has also provided a novel discussion of the interconnections and tensions between environmental and social policy. The typology discussed the different perspectives based on their different characteristics in terms of their appreciation of public modes of governance, their relationship to economic growth and their potential to drive a social-ecological transformation. It concluded that green economy is the most modest and post-growth the most transformational perspective. The typology served as a heuristic, providing clear guidance on how to assess policy from an eco-social policy perspective. It can also be helpful in identifying different political pathways of eco-social policy integration as it signals that quite opposing and antagonistic policy agendas can lie behind the same overarching terms—as is the case, for example, with green growth and its configurations of green economy and green Keynesianism. They entail quite opposite conceptions of the role of the government in steering green growth developments. It can also indicate a pathway towards sustainable policy-making, beginning with the status quo that reflects green economy/green Keynesianism, transitioning through recomposing consumption and production, and ultimately reaching a post-growth stage.

Applying this typology has resulted in empirical reconstructions of what this book considers to be the OECD's and World Bank's eco-social policy discourses. The reconstructions entail the different discursive contents that the two organisations promote in terms of the different policy perspectives of the eco-social policy typology: green growth, green economy, green Keynesianism, recomposing consumption and production and post-growth, while the perception of interconnections and tensions between the policy fields was discussed as part of the green Keynesian understandings.

One important reflection addresses the recomposing consumption and production perspective within the typology. This perspective is conceptualised as a potential transitional approach between green growth and post-growth (see also Gough, 2017). When applied to policy actors, it helps identify to what extent a policy actor may be shifting towards a post-growth perspective or whether their policy discourse emphasises green growth or post-growth. Specifically, regarding the OECD and World Bank, this framework clarifies that these are growth-driven organisations. Additionally, it highlights that the double movement dynamic is present in the recomposing consumption and production understanding. Many elements conform to a self-regulating market logic, which speaks to the empowerment of consumers to influence consumption and production patterns (first movement). However, to some degree, public modes of consumption and government procurement are also valued (countermovement).

While broadly speaking, the OECD and World Bank are characterised more by their similarities than their dissimilarities, there are still some crucial differences between their eco-social policy discourses. The OECD provides a more detailed analysis of the tensions and interconnections between environmental and social policy and communicates greater scepticism concerning the value of economic growth, while critical remarks concerning economic growth are rather marginal in the World Bank's discourse. Both organisations advance similar green growth notions. However, the OECD focuses on the knowledge economy, while the World Bank emphasises infrastructure projects. While the OECD has a more comprehensive understanding of green Keynesianism, the World Bank takes up a more comprehensive mandate concerning green economy (even though its green Keynesianism understanding should not be underestimated). In terms of recomposing consumption and production, they are quite similar. Both have strong market-based thinking while at the same time also partly appreciating communal consumption patterns. However, while the OECD also puts a certain emphasis on CBA and differentiated tax systems, the World Bank does not. As far as post-growth is concerned, the OECD has more sophisticated post-growth elements, being more critical of GDP as a measure of development and progress as well as having—at least at the margin of its policy discourse—elements of job guarantees, universal basic services and work time reduction policies.

Having illustrated empirical reconstructions of the OECD's and World Bank's eco-social policy discourses, this book further theorised the reconstructions, advancing four different but interconnected arguments that represent a new departure in terms of understanding eco-social policy dynamics with respect to global policy and global governance. The first of these argued that the eco-social policy discourses of both organisations are characterised by a double movement dynamic as conceptualised by Polanyi

(2001). Green economy represents the first movement towards a (green) self-regulating market. While the traditional conception of Polanyi's first movement entirely disregards the ecology, green economy acknowledges that it is crucial to incorporate ecological concerns into economic decision-making, although it still closely aligns with the idea of a self-regulating market, calls only for limited government intervention and still commodifies nature. Green Keynesianism, on the other hand, represents the countermovement in as much as it calls for comprehensive government intervention and a set of social protection systems. This dynamic interplay between the two movements within eco-social policy is a crucial mechanism for stabilising the contemporary capitalist mode of production in global governance. This suggests that within both organisations the first movement and the countermovement can be found "under the same roof" as Mendly (2020, p. 515) speculated in her theoretical work on the double movement in the global governance system.

The second argument assessed the OECD's and World Bank's eco-social policies in terms of their potential to drive a social-ecological transformation. The book has argued that both organisations still advocate substantial elements of "liberal globalism" while they also move towards a policy agenda of green embedded liberalism. This is also reflected in the double movement dynamic between green economy (liberal globalism) and green Keynesianism (green embedded liberalism). However, both organisations fall short of promoting more radical policy proposals that involve decommodifying nature and labour and post-growth policies that follow criteria of socio-economic democratisation, emphasising communal modes of possession, and selective economic deglobalisation while still pursuing a cooperative global governance logic.

The third argument suggested that the OECD's and World Bank's eco-social policy discourses follow the pattern of what Brand et al. (2020) call a "new critical orthodoxy". This means that they recognise that there are profound and far-reaching ecological and social crises which require some sort of large-scale transformative change. However, they then go on to propose solutions built on existing institutions and capitalist dynamics that are assumed to be capable of solving contemporary crises and problems if only they are adapted in specific ways (Brand et al., 2020). Accordingly, green Keynesianism represents a move towards a "new" capitalist mode of production in which the government reclaims its control over the market to further drive logics of capital accumulation and valorising nature, though with a more profound set of social protection systems and interventions to prevent detrimental effects. This also raises awareness of the structural conditions of the global political economy and how this global economic system, geared towards economic growth, seeks to control policy integration in a way that constructs the ecological sphere and the social sphere to be in service to the economy. Such a perspective is essential to understanding certain limitations concerning policy integration within the incumbent economic order. Eco-social policy scholarship needs to be aware that policy integration (regardless of its scale) must relate to those structuring conditions of the global political economy (Schulze Waltrup et al., 2023). This means that there will be persistent pressure from actors at various scales that benefit from such a global economic system to steer policy integration to the advantage of the incumbent system.

The fourth argument assessed the eco-social policy dynamics identified in terms of policy paradigm change. The struggle between green economy and green Keynesianism indicates a process of paradigm change or at least a struggle between third-order paradigms to become or remain incumbent. This would fall short of a fourth-order change of "general systemic logics" (Buch-Hansen & Carstensen, 2021, p. 312), which would abandon the economic-growth-based capitalist mode of production and move towards post-growth policy ambitions to radically decommodify nature and labour. Such a perspective also suggests that there is a need to reconceptualise the Polanyian countermovement. Historically, it represented policy ambitions of government intervention and social protection within a growth-based capitalist system (as illustrated in third-order paradigm change (Hall, 1993)), but this is what post-growth rejects. The latter appears to be, to a certain degree, another countermovement different from the traditional and "original" conception—in other words, a counter-countermovement. Further work should dive deeper into the conceptual relationship between post-growth movements and the countermovement that Polanyi (2001) conceptualised.

The book has particularly built on the tradition of global social policy research, which, for example, investigates policy prescriptions by IOs and how they aim to change national social policy settings (e.g. Béland & Orenstein, 2013; Deacon, 2007; Kaasch, 2015; Martens et al., 2021) and establish transnational social governance arrangements to navigate and advance their social policy interests (Kaasch & Martens, 2015). In relation to the "architecture of arguments" (Niemann et al., 2021) in global social governance, this book has assessed policy discourses of the OECD and

World Bank in terms of their application of an integrated argumentative perspective to environmental and social policy issues. In other words, the book has reconstructed the degree to which the organisations connect arguments of environmental and social policy. By researching the OECD and the World Bank, it has focused on "comprehensive IOs" (Niemann et al., 2021, p. 15) that are not characterised by a narrow and specific policy focus but speak to various issues and policy fields and then potentially link the different concerns with each other. They are most suitable for linking social policy issues with, for example, ecological ones (but other issues as well). Therefore, further studies could examine IOs with a "policy-specific concern" (Niemann et al., 2021, p. 14) that have a narrower focus within social policy and traditionally do not have units working in other policy fields, as is the case with the ILO (see Montesano et al., 2023 for a starting point). It could be expected that organisations with a narrower focus on social policy issues might face challenges in integrating environmental policy considerations due to the absence of units dedicated to environmental issues within their organisational structure.

This book has some limitations. It needs to be noted that during the time of writing, the new OECD Secretary-General Mathias Cormann, who took office in March 2021, changed the status of the working unit NAEC. Various parts of the analysis have referred to NAEC's work, in particular to their flagship report *Beyond Growth. New Approaches to Economic Challenges.* However, Cormann to some degree cancelled NAEC as an independent working group. Previously, it had been quite independent in its operation because it was closely connected to the office of former Secretary-General Angel Gurría, but Cormann transferred it to operate under the authority of the Economics Department, which meant an organisational depreciation of the alternative economic conceptions that NAEC had been exploring (Schulze Waltrup, 2024). While this suggests, at least initially, a reduction in the importance of the policy advice that NAEC produced (which contained quite striking elements of green Keynesianism), it is clearly evident how the political dynamics within the double movement and the discursive struggles to define the "rules of the game" have unfolded in this case. The move by Mathias Cormann can be interpreted as a first-movement attempt within the OECD to scale down alternative economic approaches that challenged incumbent more neoclassical economic thinking. This book therefore suggests that this move was part of the discursive reproduction and renegotiation of policy advice with its contemporary capitalist mode of production and signals the relevance of the applied theoretical framework (especially the double movement dynamic) and the different lines of argumentation employed in this study. There has been some public outcry concerning this decision, culminating in an open letter to Cormann from more than 30 prestigious economists calling on him to reverse his decision.[1]

It is important to note that not all policy elements assessed in this book from an eco-social perspective result from explicit processes of merging environmental and social policy working units at the organisational level. According to Mandelli (2022), such merging would be a crucial criterion for considering something as eco-social policy. This was done because outcomes resulting from explicit mergers are actually quite rare. However, as this book has illustrated, it is also evident that outcomes created within policy silos can include policy content that addresses the integration of environmental and social policy by focusing on transforming trade-offs into synergies, another important criterion according to Mandelli (2022). Most of the discursive elements originate from departments that are to some extent connected to an environmental policy agenda, while some also originate from social policy or inclusive growth working units. The particular case of the OECD's report *The Inequalities–Environment Nexus. Towards a People-Centred Green Transition*, which, in terms of the staff that produced the report, is an explicit merger of green and inclusive growth streams is rather an exception.

Another reflection pertains to the eco-social policy typology applied. As discussed, the typology is particularly valuable as far as the question of "how to assess policy actors from an eco-social policy perspective" is concerned, which is already a novel contribution to the literature. The typology works particularly well as a means to discover the eco-social ideas and discourses that organisations advocate. However, such an application is less well able to find out what processes these ideas emerge from only where they do exist. It also cannot find out what potential policy discourses and conflicts they possibly result from. Part of global governance and especially global social policy research focuses on exactly these dynamics between actors within IOs and reconstructs how important the engagement of certain individuals or departments is in enforcing or diffusing selected ideas (Deacon, 2013; Mahon, 2015; Martens & Jacobi, 2010). However, while not necessarily the focus of this book, there are still crucial

[1] See the following link: https://newforum.org/en/new-thinkers-sidelined-top-economists-criticize-oecd-chief-cormann/

findings regarding the process whereby eco-social policy ideas are established in the organisations. The finding concerning the double movement within both organisations is especially useful in revealing the struggle between different perspectives within those organisations indicating that there are indeed units within both that have characteristics of the countermovement and aim to curtail the scope of the first movement. The case of Cormann cancelling the working unit NAEC is illustrative of these processes, which can be interpreted as a move on the part of the first movement to curtail the countermovement. This also emphasises how a change of Secretary-General within the OECD has an impact on the diffusion and importance of organisational units that disseminate eco-social policy notions.

While this book has provided a foundation for the study of global eco-social policy, there is much more to study. This speaks to the politics of eco-social policy ideas within IOs (on the politics of eco-social policy, see Graziano, 2024) that could reconstruct how certain individuals or networks have pushed for eco-social policy integration in IOs. Future research could also study the institutional mechanisms or the path-dependency logics that impede or provoke eco-social policy changes as well as the more explicit junctures between IOs and national policy-makers in terms of eco-social policy implementation at the national level. This indicates that there is a substantial need for further eco-social policy research at the global level of IOs.

References

Béland, D., & Orenstein, M. A. (2013). International organizations as policy actors: An ideational approach. *Global Social Policy: An Interdisciplinary Journal of Public Policy and Social Development, 13*(2), 125–143. https://doi.org/10.1177/1468018113484608

Brand, U., Görg, C., & Wissen, M. (2020). Overcoming neoliberal globalization: Social-ecological transformation from a Polanyian perspective and beyond. *Globalizations, 17*(1), 161–176. https://doi.org/10.1080/14747731.2019.1644708

Buch-Hansen, H., & Carstensen, M. B. (2021). Paradigms and the political economy of ecopolitical projects: Green growth and degrowth compared. *Competition & Change, 25*(3–4), 308–327. https://doi.org/10.1177/1024529420987528

Cigna, L., Fischer, T., Abuannab, E. H., Heins, E., & Rathgeb, P. (2023). Varieties of just transition? Eco-social policy approaches at the international level. *Social Policy and Society, 22*(4), 730–746. https://doi.org/10.1017/S1474746423000192

Deacon, B. (2007). *Global social policy & governance*. Sage.

Deacon, B. (2013). *Global social policy in the making: The foundations of the social protection floor*. Policy Press.

Gough, I. (2017). *Heat, greed and human need. Climate change, capitalism and sustainable wellbeing*. Edward Elgar Publishing.

Graziano, P. (2024). The politics of the EU eco-social policies. *European Political Science, 23*(1), 27–38. https://doi.org/10.1057/s41304-023-00455-4

Hall, P. A. (1993). Policy paradigms, social learning, and the state: The case of economic policymaking in Britain. *Comparative Politics, 25*(3), 275–296. https://doi.org/10.2307/422246

Kaasch, A. (2015). *Shaping Global Health Policy. Global social policy actors and ideas about health care systems*. Palgrave Macmillan.

Kaasch, A., & Martens, K. E. (2015). *Actors and agency in global social governance*. Oxford University Press.

Köhler, G. (2016). Assessing the SDGs from the standpoint of eco-social policy: Using the SDGs subversively. *Journal of International and Comparative Social Policy, 32*(2), 149–164. https://doi.org/10.1080/21699763.2016.1198715

Lakeman, S. (2021). IOs and climate change: Toward global eco-social policy. In K. Martens, D. Niemann, & A. Kaasch (Eds.), *International organizations in global social governance* (pp. 255–273). Palgrave Macmillan.

Mahon, R. (2015). The OECD, gender, and social policy—Parallel stories. In *Actors and Agency in Global Social Governance* (pp. 81–101). Oxford University Press.

Mandelli, M. (2022). Understanding eco-social policies: A proposed definition and typology. *Transfer: European Review of Labour and Research, 28*(3), 333–348. https://doi.org/10.1177/10242589221125083

Martens, K., & Jacobi, A. (2010). *Mechanisms of OECD governance: International incentives for national policy-making?* Oxford University Press.

Martens, K., Niemann, D., & Kaasch, A. (Eds.). (2021). *International organizations in global social governance*. Palgrave Macmillan.

Mendly, D. (2020). Global governance and the double movement. *Global Governance: A Review of Multilateralism and International Organizations, 26*(3), 500–521. https://doi.org/10.1163/19426720-02603006

Montesano, F. S., Biermann, F., Kalfagianni, A., & Vijge, M. J. (2023). Can the sustainable development goals green international organisations? Sustainability integration in the International Labour Organisation. *Journal of Environmental Policy & Planning, 25*(1), 1–15. https://doi.org/10.1080/1523908X.2021.1976123

Niemann, D., Martens, K., & Kaasch, A. (2021). The architecture of arguments in global social governance: Examining populations and discourses of international organizations in social policies. In K. Martens, D. Niemann, & A. Kaasch (Eds.), *International organizations in global social governance* (pp. 3–28). Palgrave Macmillan.

Polanyi, K. (2001 [1944]). *The Great Transformation. The political and economic origins of our time*. Beacon Press.

Schulze Waltrup, R. (2024). The OECD, global challenges and contestation of the economic growth paradigm. In A. Isakova, M. Neuwinger, R. Schulze Waltrup, & O. Uraiqat (Eds.), *Constructing global challenges in world politics* (pp. 214–227). Routledge.

Schulze Waltrup, R. (2025). An eco-social policy typology: From system reproduction to transformation. *Global Social Policy, 25*(1), 17–35. https://doi.org/10.1177/14680181231205777

Schulze Waltrup, R., Moore, M., & Paulsen, T. (2023). Eco-social policy in the global political economy: Analysing shifting discourses on agricultural subsidies. *European Journal of Social Security, 25*(4), 506–522. https://doi.org/10.1177/13882627231206002

Smyth, P., & Deeming, C. (2018). Conclusion: Towards a new global social policy framework? In C. Deeming & P. Smyth (Eds.), *Reframing global social policy. Social investment for sustainable and inclusive growth* (pp. 323–342). Policy Press.

List of Analysed Documents

The list displays all documents that were analysed and quoted in Sects. 4.3 and 5.3. More documents were considered and reviewed but due to overlaps or missing relevance were not integrated into the analysis that is outlined in this book.

Hallegatte, S., Bangalore, M., Bonzanigo L., Fay M., Kane T., Narloch U., Rozenberg J., Treguer D., & Vogt-Schilb, A. (2016). *Shock Waves: Managing the Impacts of Climate Change on Poverty. Climate Change and Development Series.* Washington, D.C.: The World Bank.

OECD. (2006). *The Political Economy of Environmentally Related Taxes.* Paris: OECD Publishing.

OECD. (2010). *Consumer Policy Toolkit.* Paris: OECD Publishing.

OECD. (2011a). *Towards Green Growth.* Paris: OECD Publishing.

OECD. (2011b). *Greening Household Behaviour: The Role of Public Policy.* Paris: OECD Publishing.

OECD. (2011c). *How's Life?: Measuring well-being.* Paris: OECD Publishing.

OECD. (2014). *Green Growth and Sustainable Development Forum. Conference Summary Report. Addressing the social implications of green growth.* Paris: OECD Publishing.

OECD. (2015a). *Towards Green Growth?: Tracking Progress.* Paris: OECD Publishing.

R. Schulze Waltrup, *Eco-Social Policy in the OECD and the World Bank*, https://doi.org/10.1007/978-3-031-95429-0

OECD. (2015b). *Final NAEC Synthesis. New Approaches to Economic Challenges*. Paris: OECD Publishing.
OECD. (2015c). *Material Resources, Productivity and the Environment*. Paris: OECD Publishing.
OECD. (2015d). *Creating Incentives for Greener Products: A Policy Manual for Eastern Partnership Countries*. Paris: OECD Publishing.
OECD. (2016). *Extended Producer Responsibility: Updated Guidance for Efficient Waste Management*. Paris: OECD Publishing.
OECD. (2017a). *Investing in Climate, Investing in Growth*. Paris: OECD Publishing.
OECD. (2017b). *Tackling Environmental Problems with the Help of Behavioural Insights*. Paris: OECD Publishing.
OECD. (2018a). *Good Jobs for All in a Changing World of Work: The OECD Jobs Strategy*. Paris: OECD Publishing.
OECD. (2018b). *Opportunities for All: A Framework for Policy Action on Inclusive Growth*. Paris: OECD Publishing.
OECD. (2019a). *Accelerating Climate Action: Refocusing Policies through a Well-being Lens*. Paris: OECD Publishing.
OECD. (2019b). *Waste Management and the Circular Economy in Selected OECD Countries: Evidence from Environmental Performance Reviews*. Paris: OECD Publishing.
OECD. (2019c). *Taxing Energy Use 2019: Using Taxes for Climate Action*. Paris: OECD Publishing.
OECD. (2020a). *Beyond Growth: New Approaches to Economic Challenges*. Paris: OECD Publishing.
OECD. (2020b). *Environmental health and strengthening resilience to pandemics*. Paris: OECD Publishing.
OECD. (2020c). *Building Back Better. A Sustainable, Resilient Recovery after COVID-19*. Paris: OECD Publishing.
OECD. (2020d). *Making the Green Recovery work for jobs, income and growth*. Paris: OECD Publishing.
OECD. (2020e). *Improving resource efficiency and the circularity of economies for a greener world*. Paris: OECD Publishing.
OECD. (2020f). *Towards Sustainable Land Use: Aligning Biodiversity, Climate and Food Policies*. Paris: OECD Publishing.
OECD. (2020g). *Green budgeting and tax policy tools to support a green recovery*. Paris: OECD Publishing.
OECD. (2020h). *COVID—19 and the low-carbon transition. Impacts and possible policy responses*. Paris: OECD Publishing.

OECD. (2020i). *How's Life? 2020: Measuring Well-being*. Paris: OECD Publishing.

OECD. (2020j). *Developing Sustainable Finance Definitions and Taxonomies*. Paris: OECD Publishing.

OECD. (2021a). *The Inequality/Environment Nexus: Towards a people-centred green transition*. Paris: OECD Publishing.

OECD. (2021b). *Green Budgeting in OECD Countries*. Paris: OECD Publishing.

OECD. (2021c). *Green Budget Tagging: Introductory Guidance & Principles*. Paris: OECD Publishing.

OECD. (2021d). *Effective Carbon Rates 2021. Pricing Carbon Emissions Through Taxes and Emissions Trading*. Paris: OECD Publishing.

OECD/IEA/NEA/ITF. (2015). *Aligning Policies for a Low-carbon Economy*. Paris: OECD Publishing.

PRI & World Bank. (2020). *How Policy Makers Can Implement Reforms for a Sustainable Financial System*. Washington, D.C.: The World Bank.

World Bank. (2010). *World Development Report. Development and Climate Change*. Washington, D.C.: The World Bank.

World Bank. (2012a). *Inclusive Green Growth. The Pathway to Sustainable Development*. Washington, D.C.: The World Bank.

World Bank. (2012b). *Toward a Green, Clean and Resilient World for All. A World Bank Group Environment Strategy 2012–2022*. Washington, D.C.: The World Bank.

World Bank. (2017). *Environmental and Social Framework*. Washington, D.C.: The World Bank.

World Bank & UNEP. (2017). *Roadmap for a Sustainable Financial System*. Washington, D.C.: The World Bank

World Bank, Global Partnership for Financial Inclusion, & German Federal Ministry for Economic Cooperation and Development. (2017). *Financing for SMEs in Sustainable Global Value Chains*. Washington, D.C.: The World Bank.

World Bank, & World Resource Institute. (2019). *Integrating Green and Gray. Creating Next Generation Infrastructure*. Washington, D.C.: The World Bank.

World Bank. (2019a). *The World Bank Group Action Plan on Climate Change Adaptation and Resilience*. Washington, D.C.: The World Bank.

World Bank. (2020a). *Resilient Industries. Competitiveness in the Face of Disaster*. Washington, D.C.: The World Bank.

World Bank. (2020b). *Transformative Climate Finance: A new approach for climate finance to achieve low-carbon resilient development in developing countries*. Washington, D.C.: The World Bank.

World Bank. (2020c). *Developing a National Green Taxonomy. A World Bank Guide*. Washington, D.C.: The World Bank.

World Bank & PMR. (2021). *A Guide to Developing Domestic Carbon Crediting Mechanisms*. Washington, D.C.: The World Bank.

World Bank. (2021a). *The Economic Case for Nature. A global Earth-economy model to assess development pathways*. Washington, D.C.: The World Bank.

World Bank. (2021b). *Climate Change Action Plan 2021–2025. Supporting Green, Resilient and Inclusive Development*. Washington, D.C.: The World Bank.

World Bank. (2021c). *Circular Economy in Industrial Parks: Technologies for Competitiveness*. Washington, D.C.: The World Bank.

World Bank. (2021d). *State and Trends of Carbon Pricing 2021*. Washington, D.C.: The World Bank.

World Bank. (2021e). *Toolkits for Policymakers to Green the Financial System*. Washington, D.C.: The World Bank.

World Bank. (2022). *The Role of Extended Producer Responsibility Scheme for Packaging towards Circular Economies in APEC*. Washington, D.C.: The World Bank.

References

Agrawal, N. (2016). Inequality in India: What's the real story? https://www.weforum.org/agenda/2016/10/inequality-in-india-oxfam-explainer/

Alcott, B. (2013). Should degrowth embrace the job guarantee? *Journal of Cleaner Production, 38*, 56–60. https://doi.org/10.1016/j.jclepro.2011.06.007

Alexander, S. (2015). Basic and maximum income. In G. D'Alisa, F. Demaria, & G. Kallis (Eds.), *Degrowth. A vocabulary for a new era* (pp. 146–148). Routledge.

Allan, B. B. (2019). Paradigm and nexus: Neoclassical economics and the growth imperative in the World Bank, 1948–2000. *Review of International Political Economy, 26*(1), 183–206. https://doi.org/10.1080/09692290.2018.1543719

Andersson, J. O. (2010). Basic income from an ecological perspective. *Basic Income Studies, 4*(2). https://doi.org/10.2202/1932-0183.1180

Atzmüller, R., Aulenbacher, B., Brand, U., Décieux, F., Fischer, K., & Sauer, B. (Eds.). (2019). *Capitalism in transformation. Movements and countermovements in the 21st century*. Edward Elgar.

Bailey, D. (2015). The environmental paradox of the welfare state: The dynamics of sustainability. *New Political Economy, 20*(6), 793–811. https://doi.org/10.1080/13563467.2015.1079169

Barde, J.-P. (2007). Harnessing the political economy of environmental policy: David Pearce's contribution to OECD. *Environmental and Resource Economics, 37*(1), 33–42. https://doi.org/10.1007/s10640-007-9113-8

Bärnthaler, R., & Gough, I. (2023). Provisioning for sufficiency: Envisaging production corridors. *Sustainability: Science, Practice and Policy, 19*(1), 2218690. https://doi.org/10.1080/15487733.2023.2218690

R. Schulze Waltrup, *Eco-Social Policy in the OECD and the World Bank*, https://doi.org/10.1007/978-3-031-95429-0

Bärnthaler, R., Novy, A., & Stadelmann, B. (2020). A Polanyi-inspired perspective on social-ecological transformations of cities. *Journal of Urban Affairs*, 1–25. https://doi.org/10.1080/07352166.2020.1834404

Barry, J. (2012). *The politics of actually existing unsustainability: Human flourishing in a climate-changed, carbon-constrained world*. Oxford University Press.

Bebbington, A., Guggenheim, S., Olson, E., & Woolcock, M. (2004). Exploring social capital debates at the World Bank. *Journal of Development Studies, 40*(5), 33–64. https://doi.org/10.1080/0022038042000218134

Beça, P., & Santos, R. (2010). Measuring sustainable welfare: A new approach to the ISEW. *Ecological Economics, 69*(4), 810–819. https://doi.org/10.1016/j.ecolecon.2009.11.031

Beddoe, R., Costanza, R., Farley, J., Garza, E., Kent, J., Kubiszewski, I., et al. (2009). Overcoming systemic roadblocks to sustainability: The evolutionary redesign of worldviews, institutions, and technologies. *Proceedings of the National Academy of Science U S A, 106*(8), 2483–2489. https://doi.org/10.1073/pnas.0812570106

Béland, D., & Orenstein, M. A. (2013). International organizations as policy actors: An ideational approach. *Global Social Policy: An Interdisciplinary Journal of Public Policy and Social Development, 13*(2), 125–143. https://doi.org/10.1177/1468018113484608

Bernstein, S. (2001). *The compromise of liberal environmentalism*. Columbia University Press.

Berr, E. (2015). Sustainable development in a post Keynesian perspective: Why eco-development is relevant to post Keynesian economics. *Journal of Post Keynesian Economics, 37*(3), 459–480. https://doi.org/10.1080/01603477.2015.1000173

Birdsall, N. (2005). *The world is not flat: Inequality and injustice in our global economy*. WIDER Annual Lecture 9, Helsinki: UNU World Institute for Development Economics Research. https://www.cgdev.org/sites/default/files/archive/doc/commentary/speeches/Birdsall_WIDERpaper.pdf

Bishop, M. L., & Payne, A. (2021). The political economies of different globalizations: Theorizing reglobalization. *Globalizations, 18*(1), 1–21. https://doi.org/10.1080/14747731.2020.1779963

Blackwater, B. (2012). Two cheers for environmental Keynesianism. *Capitalism Nature Socialism, 23*(2), 51–74. https://doi.org/10.1080/10455752.2012.675232

Blyth, M. M. (2001). The transformation of the Swedish model: Economic ideas, distributional conflict, and institutional change. *World Politics, 54*(1), 1–26. https://doi.org/10.1353/wp.2001.0020

Blyth, M. M. (2002). *Great transformations. Economic ideas and institutional change in the twentieth century*. Cambridge University Press.

Bohnenberger, K. (2020). Money, vouchers, public infrastructures? A framework for sustainable welfare benefits. *Sustainability, 12*(2). https://doi.org/10.3390/su12020596

Bonal, X. (2002). Plus ça change … The World Bank global education policy and the post-Washington consensus. *International Studies in Sociology of Education, 12*(1), 3–22. https://doi.org/10.1080/09620210200200080

Borowy, I. (2017a). (Re-)thinking environment and economy: The organisation for economic co-operation and development and sustainable development. In W. Kaiser & J.-H. Meyer (Eds.), *International organizations in environmental protection. Conservation and globalization in the twentieth century* (pp. 211–240). Berghahn Books.

Borowy, I. (2017b). Negotiating environment: The making of the OECD environment committee and the polluter pays principle, 1968–1972. In M. Leimgruber & M. Schmelzer (Eds.), *The OECD and the international political economy since 1948* (pp. 311–334). Palgrave Macmillan.

Bowen, A., & Hepburn, C. (2014). Green growth: An assessment. *Oxford Review of Economic Policy, 30*(3), 407–422. https://doi.org/10.1093/oxrep/gru029

Bows, A., & Barrett, J. (2010). Cumulative emission scenarios using a consumption-based approach: A glimmer of hope? *Carbon Management, 1*(1), 161–175. https://doi.org/10.4155/cmt.10.17

Brand, U., & Wissen, M. (2018). *The limits to capitalist nature. Theorizing and overcoming the imperial mode of living*. Rowman & Littlefield.

Brand, U., Görg, C., & Wissen, M. (2020). Overcoming neoliberal globalization: Social-ecological transformation from a Polanyian perspective and beyond. *Globalizations, 17*(1), 161–176. https://doi.org/10.1080/14747731.2019.1644708

Bridgen, P. (2023). An eco-social solution to energy poverty? Substance and symbolism in the England's use of domestic energy efficiency policy to achieve social and environmental synergies, 1997–2023. *European Journal of Social Security, 25*(4), 388–407. https://doi.org/10.1177/13882627231195722

Brunnée, J., & Streck, C. (2013). The UNFCCC as a negotiation forum: Towards common but more differentiated responsibilities. *Climate Policy, 13*(5), 589–607. https://doi.org/10.1080/14693062.2013.822661

Buch-Hansen, H. (2014). Capitalist diversity and degrowth trajectories to steady-state economies. *Ecological Economics, 106*, 167–173. https://doi.org/10.1016/j.ecolecon.2014.07.030

Buch-Hansen, H. (2018). The prerequisites for a degrowth paradigm shift: Insights from critical political economy. *Ecological Economics, 146*, 157–163. https://doi.org/10.1016/j.ecolecon.2017.10.021

Buch-Hansen, H., & Carstensen, M. B. (2021). Paradigms and the political economy of ecopolitical projects: Green growth and degrowth compared. *Competition & Change, 25*(3–4), 308–327. https://doi.org/10.1177/1024529420987528

Buch-Hansen, H., & Koch, M. (2019). Degrowth through income and wealth caps? *Ecological Economics, 160*, 264–271. https://doi.org/10.1016/j.ecolecon.2019.03.001

Büchs, M. (2021a). Sustainable welfare: Independence between growth and welfare has to go both ways. *Global Social Policy, 21*(2), 323–327. https://doi.org/10.1177/14680181211019153

Büchs, M. (2021b). Sustainable welfare: How do universal basic income and universal basic services compare? *Ecological Economics, 189*, 107152. https://doi.org/10.1016/j.ecolecon.2021.107152

Büchs, M., & Koch, M. (2017). *Postgrowth and wellbeing. Challenges to sustainable welfare*. Palgrave Macmillan.

Büchs, M., & Koch, M. (2019). Challenges for the degrowth transition: The debate about wellbeing. *Futures, 105*, 155–165. https://doi.org/10.1016/j.futures.2018.09.002

Büchs, M., Ivanova, D., & Schnepf, S. V. (2021). Fairness, effectiveness, and needs satisfaction: New options for designing climate policies. *Environmental Research Letters, 16*(12). https://doi.org/10.1088/1748-9326/ac2cb1

Budowski, M., & Künzler, D. (2022). Editorial: Social sustainability: An ambiguous concept, areas of tension and ways forward. *sozialpolitik.ch, 2022*(1), 1–19. https://doi.org/10.18753/2297-8224-185

Burawoy, M. (2015). Facing an unequal world. *Current Sociology, 63*(1), 5–34. https://doi.org/10.1177/0011392114564091

Burns, S. L., & Giessen, L. (2016). Dismantling comprehensive forest bureaucracies: Direct access, the World Bank, agricultural interests, and neoliberal administrative reform of forest policy in Argentina. *Society & Natural Resources, 29*(4), 493–508. https://doi.org/10.1080/08941920.2015.1089608

Burns, S. L., Krott, M., Sayadyan, H., & Giessen, L. (2017). The World Bank improving environmental and natural resource policies: Power, deregulation, and privatization in (post-Soviet). *World Development, 92*, 215–224. https://doi.org/10.1016/j.worlddev.2016.12.030

Busch, P.-O. (2009). The OECD environmental directorate: The art of persuasion and its limitations. In F. Biermann & B. Siebenhüner (Eds.), *Managers of global change: The influence of international environmental bureaucracies* (pp. 75–99). MIT Press.

Cairney, P., & Weible, C. M. (2015). Comparing and contrasting Peter Hall's paradigms and ideas with the advocacy coalition framework. In J. Hogan & M. Howlett (Eds.), *Policy paradigms in theory and practice: Discourses, ideas and anomalies in public policy dynamics* (pp. 83–99). Palgrave Macmillan UK.

Carson, M., Burns, T. R., & Calvo, D. (2009). Introduction. In M. Carson, T. R. Burns, & D. Calvo (Eds.), *Paradigms in public policy. Theory and practice of paradigm shifts in the EU* (pp. 11–28). Internationaler Verlag der Wissenschaften.

Cassiers, I. (Ed.). (2015). *Redefining prosperity.* Routledge.

Cato, M. S. (2013). *The paradox of green Keynesianism.* Green House.

Cigna, L., Fischer, T., Abuannab, E. H., Heins, E., & Rathgeb, P. (2023). Varieties of just transition? Eco-social policy approaches at the international level. *Social Policy and Society, 22*(4), 730–746. https://doi.org/10.1017/S1474746423000192

Colby, M. E. (1989). *The evolution of paradigms of environmental management in development.* In Strategic Planning and Review Department Working Paper WPS313. World Bank.

Cömert, M. (2019). Revival of Keynesian economics or greening capitalism: "Green Keynesianism". *Sosyoekonomi, 129–144.* https://doi.org/10.17233/sosyoekonomi.2019.04.07

Cooper, G. (2008). *The origin of financial crises: Central banks, credit bubbles and the efficient market fallacy.* Harriman House.

Coote, A., & Goodwin, N. (2010). *The Great Transition: Social justice and the core economy.* NEF Working Paper 1. https://neweconomics.org/uploads/files/82c90c4bb4d6147dc3_1fm6bxppl.pdf

Coote, A., & Percy, A. (2020). *The case for universal basic services.* John Wiley & Sons.

Cosme, I., Santos, R., & O'Neill, D. W. (2017). Assessing the degrowth discourse: A review and analysis of academic degrowth policy proposals. *Journal of Cleaner Production, 149,* 321–334. https://doi.org/10.1016/j.jclepro.2017.02.016

Costanza, R., Cumberland, J. H., Daly, H., Goodland, R., Norgaard, R. B., Kubiszewski, I., & Franco, C. (2015). *An introduction to ecological economics.* CRC Press.

Cotta, B. (2024). Unpacking the eco-social perspective in European policy, politics, and polity dimensions. *European Political Science, 23*(1), 1–13. https://doi.org/10.1057/s41304-023-00453-6

Cotta, B. (2025). The eco-social aspects of the European Green Deal and the Farm to Fork. *Global Social Policy, 25*(1), 112–130. https://doi.org/10.1177/14680181241261068

Custers, P. (2010). The tasks of Keynesianism today: Green new deals as transition towards a zero growth economy? *New Political Science, 32*(2), 173–191. https://doi.org/10.1080/07393141003722032

D'Alisa, G., Deriu, M., & Federico, D. (2015). Care. In G. D'Alisa, D. Federico, & G. Kallis (Eds.), *Degrowth. A vocabulary for a new era* (pp. 63–66). Routledge.

Daly, H. E. (1992). Allocation, distribution, and scale: Towards an economics that is efficient, just, and sustainable. *Ecological Economics, 6*(3), 185–193. https://doi.org/10.1016/0921-8009(92)90024-m

Daly, H. E. (2008). A steady-state economy. In *SDC workshop 'confronting structure'—Achieving economic sustainability*. Sustainable Development Commission.

Daly, H. E., & Farley, J. (2011). *Ecological economics. Principles and applications* (3rd ed.). Island Press.

De Camillis, C., & Goralczyk, M. (2012). Towards stronger measures for sustainable consumption and production policies: Proposal of a new fiscal framework based on a life cycle approach. *The International Journal of Life Cycle Assessment, 18*(1), 263–272. https://doi.org/10.1007/s11367-012-0460-5

De Francesco, F. (2014). Transfer agents, knowledge authority, and indices of regulatory quality: A comparative analysis of the World Bank and the organisation for economic co-operation and development. *Journal of Comparative Policy Analysis: Research and Practice, 18*(4), 350–365. https://doi.org/10.1080/13876988.2014.882648

De Francesco, F., & Guaschino, E. (2019). Reframing knowledge: A comparison of OECD and World Bank discourse on public governance reform. *Policy and Society, 39*(1), 113–128. https://doi.org/10.1080/14494035.2019.1609391

Deacon, B. (2007). *Global social policy & governance*. Sage.

Deacon, B. (2013). *Global social policy in the making: The foundations of the social protection floor*. Policy Press.

Deacon, B. (2014). Toward a transformative global social policy? In A. Kaasch & P. Stubbs (Eds.), *Transformations in global and regional social policies* (pp. 201–217). Palgrave Macmillan.

Dean, M. (1999). *Governmentality. Power and rule in modern society*. Sage Publications.

Deeming, C. (2021). The struggle for social sustainability. In C. Deeming (Ed.), *The struggle for social sustainability. Moral conflicts in global social policy* (pp. 291–345). Policy Press.

Deeming, C., & Smyth, P. E. (Eds.). (2018). *Reframing global social policy. Social investment for sustainable and inclusive growth*. Policy Press.

Demaria, F., Schneider, F., Sekulova, F., & Martinez-Alier, J. (2013). What is degrowth? From an activist slogan to a social movement. *Environmental Values, 22*(2), 191–215. https://doi.org/10.3197/096327113x13581561725194

Dengler, C., & Lang, M. (2022). Commoning care: Feminist degrowth visions for a socio-ecological transformation. *Feminist Economics, 28*(1), 1–28. https://doi.org/10.1080/13545701.2021.1942511

Deutschmann, C. (2019). Karl Polanyi as a theorist of disembedded markets. In R. Atzmüller, B. Aulenbacher, U. Brand, F. Décieux, K. Fischer, & B. Sauer (Eds.), *Capitalism in transformation. Movements and countermovements in the 21st century* (pp. 60–74). Edward Elgar.

Devetter, F.-X., & Rousseau, S. (2011). Working hours and sustainable development. *Review of Social Economy, 69*(3), 333–355. https://doi.org/10.1080/00346764.2011.563507

Di Giulio, A., & Fuchs, D. (2014). Sustainable consumption corridors: Concept, objections, and responses. *GAIA—Ecological Perspectives for Science and Society, 23*(3), 184–192. https://doi.org/10.14512/gaia.23.S1.6

Di Leva, C. E. (2021). The challenge and promise at the intersection of environmental and social policies: How the World Bank established a policy framework that fully integrates environmental and social concerns. *Global Social Policy, 21*(2), 344–348. https://doi.org/10.1177/14680181211019170

Diercks, G., Larsen, H., & Steward, F. (2019). Transformative innovation policy: Addressing variety in an emerging policy paradigm. *Research Policy, 48*(4), 880–894. https://doi.org/10.1016/j.respol.2018.10.028

Dietz, R., & O'Neill, D. W. (2013). *Enough is enough: Building a sustainable economy in a world of finite resources.* Routledge.

Dittmer, K. (2013). Local currencies for purposive degrowth? A quality check of some proposals for changing money-as-usual. *Journal of Cleaner Production, 54*, 3–13. https://doi.org/10.1016/j.jclepro.2013.03.044

Dittmer, K. (2015). Community currencies. In G. D'Alisa, F. Demaria, & G. Kallis (Eds.), *Degrowth. A vocabulary for a new era* (pp. 149–151). Routledge.

Douthwaite, R. (1999). *The growth illusion. How economic growth has enriched the few, impoverished the many and endangered the planet.* Green Books.

Dresner, S., & Ekins, P. (2006). Economic instruments to improve UK home energy efficiency without negative social impacts. *Fiscal Studies, 27*(1), 47–74. https://doi.org/10.1111/j.1475-5890.2006.00027.x

Druckman, A., & Jackson, T. (2008). Household energy consumption in the UK: A highly geographically and socio-economically disaggregated model. *Energy Policy, 36*(8), 3177–3192. https://doi.org/10.1016/j.enpol.2008.03.021

Dryzek, J. S., Downes, D., Hunold, C., & Schlosberg, D. (2003). *Green states and social movements: Environmentalism in the United States, United Kingdom, Germany, & Norway.* Oxford University Press.

Duit, A., Feindt, P. H., & Meadowcroft, J. (2015). Greening leviathan: The rise of the environmental state? *Environmental Politics, 25*(1), 1–23. https://doi.org/10.1080/09644016.2015.1085218

Easterlin, R. A. (1974). Does economic growth improve the human lot? Some empirical evidence. In P. A. David & M. W. Reder (Eds.), *Nations and households in economic growth* (pp. 89–125). Academic Press.

Eckersley, R. (2021). Greening states and societies: From transitions to great transformations. *Environmental Politics, 30*(1–2), 245–265. https://doi.org/10.1080/09644016.2020.1810890

Eitinger, L., & Wade, R. H. (2024). US 'hegemony'in the World Bank. In A. Vetterlein & T. Schmidtke (Eds.), *The Elgar companion to the World Bank* (pp. 118–128). Edward Elgar Publishing.

Elgin, D. (1982). *Voluntary simplicity: An ecological lifestyle that promotes personal and social renewal.* Bantam Books.

Eskelinen, T. (2015). Possibilities and limits of green Keynesianism. In K. Borgnäs, T. Eskelinen, J. Perkiö, & R. Warlenius (Eds.), *The politics of ecosocialism. Transforming welfare* (pp. 101–115). Routledge.

Fairclough, N. (2010). *Critical discourse analysis* (2nd ed.). Longman.

Falkner, R. (2008). *Business power and conflict in international environmental politics*. Palgrave Macmillan.

Falkner, R. (2016). The Paris agreement and the new logic of international climate politics. *International Affairs, 92*(5), 1107–1125. https://doi.org/10.1111/1468-2346.12708

Fanning, A. L., O'Neill, D. W., & Büchs, M. (2020). Provisioning systems for a good life within planetary boundaries. *Global Environmental Change, 64*. https://doi.org/10.1016/j.gloenvcha.2020.102135

Fanning, A. L., O'Neill, D. W., Hickel, J., & Roux, N. (2022). The social shortfall and ecological overshoot of nations. *Nature Sustainability, 5*(1), 26–36. https://doi.org/10.1038/s41893-021-00799-z

Fawcett, T., & Parag, Y. (2010). An introduction to personal carbon trading. *Climate Policy, 10*(4), 329–338. https://doi.org/10.3763/cpol.2010.0649

Fergusson, R. (2021). International organizations' involvement in youth unemployment as a global policy field, and the global financial crisis. In D. Niemann, K. Martens, & A. Kaasch (Eds.), *International organizations in global social governance* (pp. 31–56). Palgrave Macmillan.

Fergusson, R., & Yeates, N. (2013). Business, as usual: The policy priorities of the World Bank's discourses on youth unemployment, and the global financial crisis. *Journal of International and Comparative Social Policy, 29*(1), 64–78. https://doi.org/10.1080/21699763.2013.803998

Fioretos, O., & Heldt, E. C. (2019). Legacies and innovations in global economic governance since Bretton woods. *Review of International Political Economy, 26*(6), 1089–1111. https://doi.org/10.1080/09692290.2019.1635513

Fitzpatrick, T. (2011). Challenges for social policy. In T. Fitzpatrick (Ed.), *Understanding the environment and social policy* (pp. 61–90). The Policy Press.

Foucault, M. (1991). Governmentality. In G. Burchell, C. Gordon, & P. Miller (Eds.), *The Foucault effect. Studies in governmentality* (pp. 87–104). Harvester Wheatsheaf.

Foundational Economy Collective. (2018). *Foundational economy. The infrastructure of everyday life*. Manchester University Press.

Franzen, A. (2003). Environmental attitudes in international comparison: An analysis of the ISSP surveys 1993 and 2000. *Social Science Quarterly, 84*(2), 297–308. https://doi.org/10.1111/1540-6237.8402005

Fuchs, D. (2007). *Business power in global governance*. Lynne Rienner.

Fuchs, D. (2013). Sustainable consumption. In R. Falkner (Ed.), *The handbook of global climate and environment policy* (pp. 215–230). John Wiley & Sons, Ltd.

Fuchs, D., & Lorek, S. (2001). An inquiry into the impact of globalization on the potential for 'sustainable consumption'. In *Households. ProSus Report 2/01*. ProSus.

Fuchs, D., Sahakian, M., Gumbert, T., Di Giulio, A., Maniates, M., Lorek, S., & Graf, A. (2021). *Consumption corridors. Living a good life within sustainable limits*. Routledge.

Fukuda-Parr, S. (2019). Keeping out extreme inequality from the SDG agenda—The politics of indicators. *Global Policy, 10*(S1), 61–69. https://doi.org/10.1111/1758-5899.12602

Gerber, J.-F., & Raina, R. S. (2018). Post-growth in the global South? Some reflections from India and Bhutan. *Ecological Economics, 150*, 353–358. https://doi.org/10.1016/j.ecolecon.2018.02.020

Glaser, B., & Strauss, A. L. (1967). *The discovery of grounded theory: Strategies for qualitative research*. Aldine de Gruyter.

Goldman, M. (2001). Constructing an environmental state: Eco-governmentality and other transnational practices of a "green" World Bank. *Social Problems, 48*(4), 499–523. https://doi.org/10.1525/sp.2001.48.4.499

Goldstein, J., & Tyfield, D. (2017). Green Keynesianism: Bringing the entrepreneurial state back in(to question)? *Science as Culture, 27*(1), 74–97. https://doi.org/10.1080/09505431.2017.1346598

Goodwin, G. (2018). Rethinking the double movement: Expanding the frontiers of Polanyian analysis in the Global South. *Development and Change, 49*(5), 1268–1290. https://doi.org/10.1111/dech.12419

Gore, C. (2000). The rise and fall of the Washington consensus as a paradigm for developing countries. *World Development, 28*(5), 789–804. https://doi.org/10.1016/s0305-750x(99)00160-6

Gough, I. (2013). Climate change, social policy, and global governance. *Journal of International and Comparative Social Policy, 29*(3), 185–203. https://doi.org/10.1080/21699763.2013.852128

Gough, I. (2016). Welfare states and environmental states: A comparative analysis. *Environmental Politics, 25*(1), 24–47. https://doi.org/10.1080/09644016.2015.1074382

Gough, I. (2017). *Heat, greed and human need. Climate change, capitalism and sustainable wellbeing*. Edward Elgar Publishing.

Gough, I., & Meadowcroft, J. (2011). Decarbonizing the welfare state. In J. S. Dryzek, R. B. Norgaard, & D. Schlosberg (Eds.), *The Oxford handbook of climate change and society* (pp. 490–503). Oxford University Press.

Gough, I., & Therborn, G. (2010). The global future of welfare states. In F. G. Castles, S. Leibfried, J. Lewis, H. Obinger, & C. Pierson (Eds.), *The Oxford handbook of the welfare state* (pp. 703–720). Oxford University Press.

Graziano, P. (2024). The politics of the EU eco-social policies. *European Political Science, 23*(1), 27–38. https://doi.org/10.1057/s41304-023-00455-4

Greening, L. A., Greene, D. L., & Difiglio, C. (2000). Energy efficiency and consumption—The rebound effect—A survey. *Energy Policy, 28*(6–7), 389–401. https://doi.org/10.1016/S0301-4215(00)00021-5

Greer, S., Elliott, H., & Oliver, R. (2020). Differences that matter: Overcoming methodological nationalism in comparative social policy research. In *Theory and methods in comparative policy analysis studies* (pp. 267–288). Routledge.

Haas, W., Krausmann, F., Wiedenhofer, D., & Heinz, M. (2015). How circular is the global economy?: An assessment of material flows, waste production, and recycling in the European Union and the world in 2005. *Journal of Industrial Ecology, 19*(5), 765–777. https://doi.org/10.1111/jiec.12244

Hajer, M. A. (1995). *The politics of environmental discourse: Ecological modernization and the policy process.* Oxford University Press.

Hajer, M. A. (2009). Ecological modernisation as cultural politics. In A. P. J. Mol, D. A. Sonnenfeld, & G. Spaargaren (Eds.), *The ecological modernisation reader. Environmental reform in theory and practice* (pp. 80–100). Routledge.

Hall, P. A. (1993). Policy paradigms, social learning, and the state: The case of economic policymaking in Britain. *Comparative Politics, 25*(3), 275–296. https://doi.org/10.2307/422246

Hall, A. (2007). Social policies in the World Bank. *Global Social Policy, 7*(2), 151–175. https://doi.org/10.1177/1468018107078160

Hallegatte, S., Bangalore, M., Bonzanigo, L., Fay, M., Kane, T., Narloch, U., Rozenberg, J., Treguer, D., & Vogt-Schilb, A. (2016). *Shock waves: Managing the impacts of climate change on poverty. Climate change and development series.* The World Bank.

Hammer, P. J. (2013). *Change and continuity at the World Bank: Reforming paradoxes of economic development.* Edward Elgar.

Harris, J. M. (2013). *Green Keynesianism: Beyond standard growth paradigms.* GDAE Working Papers 13–02, GDAE, Medford and Somerville, MA: Tufts University.

Heneghan, M. (2021). International organizations and the global social governance of pensions. In K. Martens, D. Niemann, & A. Kaasch (Eds.), *International organizations in global social governance* (pp. 113–135). Palgrave Macmillan.

Hertwich, E. G. (2005). Consumption and the rebound effect: An industrial ecology perspective. *Journal of Industrial Ecology, 9*(1–2), 85–98. https://doi.org/10.1162/1088198054084635

Hickel, J. (2020). What does degrowth mean? A few points of clarification. *Globalizations*, 1–7. https://doi.org/10.1080/14747731.2020.1812222

Hickel, J., & Hallegatte, S. (2022). Can we live within environmental limits and still reduce poverty? Degrowth or decoupling? *Development Policy Review, 40*(1). https://doi.org/10.1111/dpr.12584

Hickel, J., & Kallis, G. (2020). Is green growth possible? *New Political Economy, 25*(4), 469–486. https://doi.org/10.1080/13563467.2019.1598964

Hickel, J., Dorninger, C., Wieland, H., & Suwandi, I. (2022). Imperialist appropriation in the world economy: Drain from the global south through unequal exchange, 1990–2015. *Global Environmental Change, 73*, 102467. https://doi.org/10.1016/j.gloenvcha.2022.102467

Hirsch, F. (2005). *Social limits to growth*. Routledge.

Hirvilammi, T. (2020). The virtuous circle of sustainable welfare as a transformative policy idea. *Sustainability, 12*(1), 391. https://doi.org/10.3390/su12010391

Hirvilammi, T., & Helne, T. (2014). Changing paradigms: A sketch for sustainable wellbeing and ecosocial policy. *Sustainability, 6*(4), 2160–2175. https://doi.org/10.3390/su6042160

Hirvilammi, T., & Koch, M. (2020). Sustainable welfare beyond growth. *Sustainability, 12*(5). https://doi.org/10.3390/su12051824

Hirvilammi, T., Peltomaa, J., Pihlajamaa, M., & Tiilikainen, S. (2023). Towards an eco-welfare state: Enabling factors for transformative eco-social initiatives. *European Journal of Social Security, 25*(4), 445–463. https://doi.org/10.1177/13882627231195724

Hood, C. (1991). A public management for all seasons? *Public Administration, 69*(1), 3–19. https://doi.org/10.1111/j.1467-9299.1991.tb00779.x

Hubacek, K., Baiocchi, G., Feng, K., Muñoz Castillo, R., Sun, L., & Xue, J. (2017). Global carbon inequality. *Energy, Ecology and Environment, 2*(6), 361–369. https://doi.org/10.1007/s40974-017-0072-9

Huber, J. (2004). *New technologies and environmental innovation*. Edward Elgar.

Huber, J., & Robertson, J. (2000). *Creating new money. A monetary reform for the information age*. New Economics Foundation.

Hynes, W. (2021). A systemic socio-ecological recovery from Covid-19. *Global Social Policy, 21*(2), 335–338. https://doi.org/10.1177/14680181211019166

Hynes, W., Lees, M., & Müller, J. M. (Eds.). (2020). *Systemic thinking for policy making. The potential of systems analysis for addressing global policy challenges in the 21st century*. OECD Publishing.

Jackson, T. (2005). *Motivating sustainable consumption. A review of evidence on consumer and behavioural change*. Report to the Sustainable Development Research Network.

Jackson, T. (2009). *Prosperity without growth: Economics for a finite planet*. Earthscan.

Jackson, T., & Webster, R. (2018). Limits to growth revisited. In C. Deeming & P. Smyth (Eds.), *Reframing global social policy. Social investment for sustainable and inclusive growth* (pp. 295–322). Policy Press.

Jacobs, M. (1991). *The green economy: Environment, sustainable development and the politics of the future*. Pluto Press.

Jacobs, M. (2013). Green growth: Economic theory and political discourse. In R. Falkner (Ed.), *Handbook of global climate and environmental policy* (pp. 197–214). Wiley Blackwell.

Jenson, J. (2010). Diffusing ideas for after neoliberalism. *Global Social Policy, 10*(1), 59–84. https://doi.org/10.1177/1468018109354813

Johanisova, N., & Wolf, S. (2012). Economic democracy: A path for the future? *Futures, 44*(6), 562–570. https://doi.org/10.1016/j.futures.2012.03.017

John, P., Smith, G., & Stoker, G. (2009). Nudge nudge, think think: Two strategies for changing civic behaviour. *The Political Quarterly, 80*(3), 361–370. https://doi.org/10.1111/j.1467-923X.2009.02001.x

Jørgensen, M. W., & Philips, L. (2002). *Discourse analysis as theory and method.* Sage Publications.

Kaasch, A. (2015). *Shaping Global Health Policy. Global social policy actors and ideas about health care systems.* Palgrave Macmillan.

Kaasch, A. (2019). Introduction to a special issue in memory of Bob Deacon: Where do we stand in global social policy studies? *Global Social Policy, 19*(1–2), 4–20. https://doi.org/10.1177/1468018119847336

Kaasch, A., & Martens, K. E. (2015). *Actors and agency in global social governance.* Oxford University Press.

Kaasch, A., & Schulze Waltrup, R. (2021). Introduction: Global eco-social policy: Contestation within an emerging policy era? *Global Social Policy, 21*(2), 319–322. https://doi.org/10.1177/14680181211019152

Kallis, G. (2011). In defence of degrowth. *Ecological Economics, 70*(5), 873–880. https://doi.org/10.1016/j.ecolecon.2010.12.007

Kallis, G. (2017). Radical dematerialization and degrowth. *Philosophical Transactions of the Royal Society A: Mathematical, Physical and Engineering Sciences, 375*(2095). https://doi.org/10.1098/rsta.2016.0383

Kallis, G. (2018). *Degrowth.* Agenda Publishing.

Kallis, G., Kalush, M., O'Flynn, H., Rossiter, J., & Ashford, N. (2013). 'Friday off': Reducing working hours in Europe. *Sustainability, 5*(4), 1545–1567. https://doi.org/10.3390/su5041545

Kallis, G., Kostakis, V., Lange, S., Muraca, B., Paulson, S., & Schmelzer, M. (2018). Research on degrowth. *Annual Review of Environment and Resources, 43*, 291–316. https://doi.org/10.1146/annurev-environ-102017-025941

Kapur, D., Lewis, J. P., & Webb, R. (1997). *The World Bank. Its first half century.* The World Bank.

Kasser, T. (2017). Living both well and sustainably: A review of the literature, with some reflections on future research, interventions and policy. *Philosophical Transactions of the Royal Society A: Mathematical, Physical and Engineering Sciences, 375*(2095). https://doi.org/10.1098/rsta.2016.0369

Kerschner, C. (2010). Economic de-growth vs. steady-state economy. *Journal of Cleaner Production, 18*(6), 544–551. https://doi.org/10.1016/j.jclepro.2009.10.019

Keynes, J. M. (1936). *The general theory of employment, interest and money*. Harcourt, Brace & World, Inc.

King, J. E. (2008). Post-Keynesian economics. In S. Durlauf & L. E. Blume (Eds.), *The new Palgrave dictionary of economics* 2nd ed. (online). Palgrave Macmillan.

Knight, K. W., Schor, J. B., & Jorgenson, A. K. (2017). Wealth inequality and carbon emissions in high-income countries. *Social Currents, 4*(5), 403–412. https://doi.org/10.1177/2329496517704872

Koch, M. (2020). The state in the transformation to a sustainable postgrowth economy. *Environmental Politics, 29*(1), 115–133. https://doi.org/10.1080/09644016.2019.1684738

Koch, M. (2022). Social policy without growth: Moving towards sustainable welfare states. *Social Policy and Society, 21*(3), 447–459. https://doi.org/10.1017/S1474746421000361

Koch, M., & Fritz, M. (2014). Building the eco-social state: Do welfare regimes matter? *Journal of Social Policy, 43*(4), 679–703. https://doi.org/10.1017/s004727941400035x

Köhler, G. (2016). Assessing the SDGs from the standpoint of eco-social policy: Using the SDGs subversively. *Journal of International and Comparative Social Policy, 32*(2), 149–164. https://doi.org/10.1080/21699763.2016.1198715

Korpi, W. (1983). *The democratic class struggle*. Routledge and Keagan Paul.

Korpi, W. (1985). Economic growth and the welfare state: Leaky bucket or irrigation system. *European Sociological Review, 1*(2), 97–118.

Krause, D. (2021). Uncovering different degrees of eco-social policy. *Global Social Policy, 21*(2), 332–334. https://doi.org/10.1177/14680181211019165

Krause, D., Stevis, D., Hujo, K., & Morena, E. (2022). Just transitions for a new eco-social contract: Analysing the relations between welfare regimes and transition pathways. *Transfer, 28*(3), 367–382. https://doi.org/10.1177/10242589221127838

Kreinin, H. (2020). *Typologies of 'just transitions': Towards social-ecological transformation*. Ecological Economic Papers, 35., WU Vienna University of Economics and Business, Vienna.

Kubiszewski, I., Costanza, R., Franco, C., Lawn, P., Talberth, J., Jackson, T., & Aylmer, C. (2013). Beyond GDP: Measuring and achieving global genuine progress. *Ecological Economics, 93*, 57–68. https://doi.org/10.1016/j.ecolecon.2013.04.019

Kuhnle, S., & Sander, A. (2010). The emergence of the western welfare state. In F. G. Castles, S. Leibfried, J. Lewis, H. Obinger, & C. Pierson (Eds.), *The Oxford handbook of the welfare state* (pp. 61–80). Oxford University Press.

Lakeman, S. (2021). IOs and climate change: Toward global eco-social policy. In K. Martens, D. Niemann, & A. Kaasch (Eds.), *International organizations in global social governance* (pp. 255–273). Palgrave Macmillan.

Lamb, W. F., Antal, M., Bohnenberger, K., Brand-Correa, L. I., Müller-Hansen, F., Jakob, M., et al. (2020). What are the social outcomes of climate policies? A systematic map and review of the ex-post literature. *Environmental Research Letters, 15*(11). https://doi.org/10.1088/1748-9326/abc11f

Lanoie, P., Patry, M., & Lajeunesse, R. (2008). Environmental regulation and productivity: Testing the Porter hypothesis. *Journal of Productivity Analysis, 30*(2), 121–128. https://doi.org/10.1007/s11123-008-0108-4

Laruffa, F., & Nullmeier, F. (2025). A model to follow? The EU and global eco-social policy. *Global Social Policy, 25*(1), 151–169. https://doi.org/10.1177/14680181241246935

Latouche, S. (2009). *Farewell to growth*. Polity.

Lebel, L., & Lorek, S. (2008). Enabling sustainable production–consumption systems. *Annual Review of Environment and Resources, 33*(1), 241–275. https://doi.org/10.1146/annurev.environ.33.022007.145734

Lechner, F. J., & Boli, J. (2005). *World culture: Origins and consequences*. Blackwell.

Lehtonen, M. (2007). Environmental policy integration through OECD peer reviews: Integrating the economy with the environment or the environment with the economy? *Environmental Politics, 16*(1), 15–35. https://doi.org/10.1080/09644010601073432

Lietaert, M. (2010). Cohousing's relevance to degrowth theories. *Journal of Cleaner Production, 18*(6), 576–580. https://doi.org/10.1016/j.jclepro.2009.11.016

Lifset, R. J. (1993). Take it back: Extended producer responsibility as a form of incentive-based environmental policy. *Journal of Resource Management and Technology, 21*(4), 163–175.

Lindellee, J., Alkan Olsson, J., & Koch, M. (2021). Operationalizing sustainable welfare and co-developing eco-social policies by prioritizing human needs. *Global Social Policy, 21*(2), 328–331. https://doi.org/10.1177/14680181211019164

Liobikienė, G., & Dagiliūtė, R. (2016). The relationship between economic and carbon footprint changes in EU: The achievements of the EU sustainable consumption and production policy implementation. *Environmental Science & Policy, 61*, 204–211. https://doi.org/10.1016/j.envsci.2016.04.017

Lipschutz, R. D., & McKendry, C. (2011). In J. S. Dryzek, R. B. Norgaard, & D. Schlosberg (Eds.), *Social movements and global civil society* (pp. 369–383). Oxford University Press.

Long, B. L. (2000). *International environmental issues and the OECD 1950–2000. An historical perspective*. OECD.

Lorek, S., & Fuchs, D. (2013). Strong sustainable consumption governance—Precondition for a degrowth path? *Journal of Cleaner Production, 38*, 36–43. https://doi.org/10.1016/j.jclepro.2011.08.008

MacMillan, M., Hoban, T. J., Clifford, W. B., & Brant, M. R. (1997). Social and demographic influences on environmental attitudes. *Journal of Rural Social Sciences, 13*(1), 89–107.

Mahon, R. (2009). The OECD'S discourse on the reconciliation of work and family life. *Global Social Policy, 9*(2), 183–204. https://doi.org/10.1177/1468018109104625

Mahon, R. (2010). After neo-liberalism? The OECD, the World Bank and the child. *Global Social Policy, 10*(2), 172–192. https://doi.org/10.1177/1468018110366615

Mahon, R. (2015). The OECD, gender, and social policy—Parallel stories. In *Actors and Agency in Global Social Governance* (pp. 81–101). Oxford University Press.

Mahon, R. (2019). Broadening the social investment agenda: The OECD, the World Bank and inclusive growth. *Global Social Policy, 19*(1–2), 121–138. https://doi.org/10.1177/1468018119826404

Mahon, R., & McBride, S. (2009). Standardizing and disseminating knowledge: The role of the OECD in global governance. *European Political Science Review, 1*(1), 83–101. https://doi.org/10.1017/S1755773909000058

Mandelli, M. (2022). Understanding eco-social policies: A proposed definition and typology. *Transfer: European Review of Labour and Research, 28*(3), 333–348. https://doi.org/10.1177/10242589221125083

Mandelli, M. (2025). The comparative politics of just transition policies: Building green–red winning coalitions in Spain and Ireland. *Journal of European Public Policy*, 1–26. https://doi.org/10.1080/13501763.2024.2446364

Markkanen, S., & Anger-Kraavi, A. (2019). Social impacts of climate change mitigation policies and their implications for inequality. *Climate Policy, 19*(7), 827–844. https://doi.org/10.1080/14693062.2019.1596873

Marschinski, R., & Behrle, S. (2009). The World Bank: Making the business case for the environment. In F. Biermann & B. Siebenhüner (Eds.), *Managers of global change: The influence of international environmental bureaucracies* (pp. 101–142). MIT Press.

Martens, K., & Jacobi, A. (2010). *Mechanisms of OECD governance: International incentives for national policy-making?* Oxford University Press.

Martens, K., Niemann, D., & Kaasch, A. (Eds.). (2021). *International organizations in global social governance.* Palgrave Macmillan.

Mayring, P. (2015). *Qualitative Inhaltsanalyse* [Qualitative content analysis] (12th ed.). Beltz.

McBride, S., & Williams, R. A. (2001). Globalization, the restructuring of labour markets and policy convergence. *Global Social Policy, 1*(3), 281–309. https://doi.org/10.1177/146801810100100302

Meadowcroft, J. (2005). From welfare state to ecostate. In R. Eckersely & J. Barry (Eds.), *The state and the global ecological crisis* (pp. 3–23). The MIT Press.

Mendly, D. (2020). Global governance and the double movement. *Global Governance: A Review of Multilateralism and International Organizations, 26*(3), 500–521. https://doi.org/10.1163/19426720-02603006

Merlingen, M. (2003). Governmentality. Towards a Foucauldian framework for the study of IGOs. *Cooperation and Conflict: Journal of the Nordic International Studies Association, 38*(4), 361–384. https://doi.org/10.1177/0010836703384002

Meyer, J. W. (2009). Reflections: Institutional theory and world society. In G. S. Drori & G. Krücken (Eds.), *World society: The writings of John W. Meyer* (pp. 36–63). Oxford University Press.

Meyer, J. W., Boli, J., & Thomas, G. M. (1987). Ontology and rationalization in the western cultural account. In G. M. Thomas, J. W. Meyer, F. O. Ramirez, & J. Boli (Eds.), *Institutional structure. Constituting state, society, and the individual* (pp. 12–37). Sage.

Meyer, J. W., Boli, J., Thomas, G. M., & Ramirez, F. O. (1997). World society and the nation-state. *American Journal of Sociology, 103*(1), 144–181. https://doi.org/10.1086/231174

Miller, P., & Rose, N. (2008). *Governing the present. Administering economic, social and personal life*. Polity Press.

Millward-Hopkins, J., & Oswald, Y. (2021). 'Fair' inequality, consumption and climate mitigation. *Environmental Research Letters, 16*(3). https://doi.org/10.1088/1748-9326/abe14f

Mol, A. P. J., & Sonnenfeld, D. A. (2000). Ecological modernisation around the world: An introduction. *Environmental Politics, 9*(1), 1–14. https://doi.org/10.1080/09644010008414510

Mol, A. P. J., & Spaargaren, G. (2000). Ecological modernisation theory in debate: A review. *Environmental Politics, 9*(1), 17–49. https://doi.org/10.1080/09644010008414511

Mol, A. P. J., Sonnenfeld, D. A., & Spaargaren, G. (Eds.). (2009). *The ecological mondernisation reader. Environmental reform in theory and practice*. Routledge.

Montesano, F. S., Biermann, F., Kalfagianni, A., & Vijge, M. J. (2023). Can the sustainable development goals green international organisations? Sustainability integration in the International Labour Organisation. *Journal of Environmental Policy & Planning, 25*(1), 1–15. https://doi.org/10.1080/1523908X.2021.1976123

Moser, C., & Dani, A. A. (Eds.). (2008). *Assets, livelihoods and social policy*. The World Bank.

Muraca, B. (2012). Towards a fair degrowth society: Justice and the right to a 'good life' beyond growth. *Futures, 44*(6), 535–545. https://doi.org/10.1016/j.futures.2012.03.014

Newell, P., & Paterson, M. (2010). *Climate capitalism. Global warming and the transformation of the global economy*. Cambridge University Press.

Niemann, D., & Martens, K. (2018). Soft governance by hard fact? The OECD as a knowledge broker in education policy. *Global Social Policy, 18*(3), 267–283. https://doi.org/10.1177/1468018118794076

Niemann, D., Martens, K., & Kaasch, A. (2021). The architecture of arguments in global social governance: Examining populations and discourses of international organizations in social policies. In K. Martens, D. Niemann, & A. Kaasch (Eds.), *International organizations in global social governance* (pp. 3–28). Palgrave Macmillan.

Novy, A. (2022). The political trilemma of contemporary social-ecological transformation—Lessons from Karl Polanyi's *The Great Transformation. Globalizations, 19*(1), 59–80. https://doi.org/10.1080/14747731.2020.1850073

Nussbaum, M. (2003). Capabilities as fundamental entitlements: Sen and social justice. *Feminist Economics, 9*(2–3), 33–59. https://doi.org/10.1080/1354570022000077926

Nyfors, T., Linnanen, L., Nissinen, A., Seppälä, J., Saarinen, M., Regina, K., et al. (2020). Ecological sufficiency in climate policy: Towards policies for recomposing consumption. *Futura, 3.*

O'Neill, K. (2009). *The environment and international relations.* Cambridge University Press.

O'Neill, D. W. (2015). The proximity of nations to a socially sustainable steady-state economy. *Journal of Cleaner Production, 108*, 1213–1231. https://doi.org/10.1016/j.jclepro.2015.07.116

O'Neill, D. W., Fanning, A. L., Lamb, W. F., & Steinberger, J. K. (2018). A good life for all within planetary boundaries. *Nature Sustainability, 1*(2), 88–95. https://doi.org/10.1038/s41893-018-0021-4

OECD. (1981). *The crisis of the welfare state.* OECD Publishing.

OECD. (1994). *New orientations in social policy.* OECD Publishing.

OECD. (2006). *The political economy of environmentally related taxes.* OECD Publishing.

OECD. (2008). Growing unequal. In *Income distribution and poverty in OECD countries.* OECD Publishing.

OECD. (2010). *Consumer policy toolkit.* OECD Publishing.

OECD. (2011a). *Towards green growth.* OECD Publishing.

OECD. (2011b). *Greening household behaviour: The role of public policy.* OECD Publishing.

OECD. (2011c). *How's life? Measuring well-being.* OECD Publishing.

OECD. (2011d). *Divided we stand. Why inequality keeps growing.* OECD Publishing.

OECD. (2014). *Green growth and sustainable development forum.* Conference Summary Report. Addressing the social implications of green growth. OECD Publishing.

OECD. (2015a). *Towards green growth? Tracking progress.* OECD Publishing.

OECD. (2015b). *Final NAEC synthesis. New approaches to economic challenges.* OECD Publishing.

OECD. (2015c). *Material resources, productivity and the environment.* OECD Publishing.

OECD. (2015d). *Creating incentives for greener products: A policy manual for eastern partnership countries.* OECD Publishing.

OECD. (2015e). *In it together. Why less inequality benefits all.* OECD Publishing.

OECD. (2016). *Extended producer responsibility: Updated guidance for efficient waste management.* OECD Publishing.

OECD. (2017a). *Investing in climate, investing in growth.* OECD Publishing.

OECD. (2017b). *Tackling environmental problems with the help of behavioural insights.* OECD Publishing.

OECD. (2018a). *Good jobs for all in a changing world of work: The OECD jobs strategy.* OECD Publishing.

OECD. (2018b). *Opportunities for all: A framework for policy action on inclusive growth.* OECD Publishing.

OECD. (2019a). *Accelerating climate action: Refocusing policies through a well-being lens.* OECD Publishing.

OECD. (2019b). *Waste management and the circular economy in selected OECD countries: Evidence from environmental performance reviews.* OECD Publishing.

OECD. (2019c). *Taxing energy use 2019: Using taxes for climate action.* OECD Publishing.

OECD. (2020a). *Beyond growth: New approaches to economic challenges.* OECD Publishing.

OECD. (2020b). *Environmental health and strengthening resilience to pandemics.* OECD Publishing.

OECD. (2020c). *Building back better. A sustainable, resilient recovery after COVID-19.* OECD Publishing.

OECD. (2020d). *Making the green recovery work for jobs, income and growth.* OECD Publishing.

OECD. (2020e). *Improving resource efficiency and the circularity of economies for a greener world.* OECD Publishing.

OECD. (2020f). *Towards sustainable land use: Aligning biodiversity, climate and food policies.* OECD Publishing.

OECD. (2020g). *Green budgeting and tax policy tools to support a green recovery.* OECD Publishing.

OECD. (2020h). *COVID–19 and the low-carbon transition. Impacts and possible policy responses.* OECD Publishing.

OECD. (2020i). *How's life? 2020: Measuring well-being.* OECD Publishing.

OECD. (2020j). *Developing sustainable finance definitions and taxonomies.* OECD Publishing.

OECD. (2021a). *The inequality/environment nexus: Towards a people-centred green transition.* OECD Publishing.

OECD. (2021b). *Green budgeting in OECD countries*. OECD Publishing.

OECD. (2021c). *Green budget tagging: Introductory guidance & principles*. OECD Publishing.

OECD. (2021d). *Effective carbon rates 2021. Pricing carbon emissions through taxes and emissions trading*. OECD Publishing.

OECD. (2023). *OECD economic outlook, interim report March 2023: A fragile recovery*. OECD Publishing.

OECD, IEA, NEA and ITF. (2015). *Aligning policies for a low-carbon economy*. OECD Publishing.

Orenstein, M. A. (2008). *Privatizing pensions*. Princeton University Press.

Orenstein, M. A. (2011). Pension privatization in crisis: Death or rebirth of a global policy trend? *International Social Security Review, 64*(3), 65–80. https://doi.org/10.1111/j.1468-246X.2011.01403.x

Osborne, D., & Gaebler, T. (1993). *Reinventing the government. How the entrepreneurial Spirit is transforming the public sector*. Plume.

Oswald, Y., Owen, A., & Steinberger, J. K. (2020). Large inequality in international and intranational energy footprints between income groups and across consumption categories. *Nature Energy, 5*(3), 231–239. https://doi.org/10.1038/s41560-020-0579-8

Patomäki, H. (2014). On the dialectics of global governance in the twenty-first century: A Polanyian double movement? *Globalizations, 11*(5), 733–750. https://doi.org/10.1080/14747731.2014.981079

Phillips, N., Lawrence, T. B., & Hardy, C. (2004). Discourse and institutions. *Academy of Management Review, 29*(4), 635–652. https://doi.org/10.2307/20159075

Pitkänen, A., Von Wright, T., Kaseva, J., & Kahiluoto, H. (2022). Distributional fairness of personal carbon trading. *Ecological Economics, 201*, 107587. https://doi.org/10.1016/j.ecolecon.2022.107587

Pizzigati, S. (2004). *Greed and good: Understanding and overcoming the inequality that limits our lives*. The Apex Press.

Polanyi, K. (2001 [1944]). *The Great Transformation. The political and economic origins of our time*. Beacon Press.

Porter, M., & van der Linde, C. (1995). Toward a new conception of the environment—Competitiveness relationship. *The Journal of Economic Perspectives, 9*(4), 97–118. https://doi.org/10.1257/jep.9.4.97

Princen, S., & Van Esch, F. (2015). Paradigm formation and paradigm change in the EU's stability and growth pact. *European Political Science Review, 8*(3), 355–375. https://doi.org/10.1017/s1755773915000089

Raftery, A. E., Zimmer, A., Frierson, D. M. W., Startz, R., & Liu, P. (2017). Less than 2 degrees C warming by 2100 unlikely. *Nature Climate Change, 7*, 637–641. https://doi.org/10.1038/nclimate3352

Rao, S., Pachauri, S., Dentener, F., Kinney, P., Klimont, Z., Riahi, K., & Schoepp, W. (2013). Better air for better health: Forging synergies in policies for energy access, climate change and air pollution. *Global Environmental Change, 23*(5), 1122–1130. https://doi.org/10.1016/j.gloenvcha.2013.05.003

Raworth, K. (2017). *Doughnut economics*. Random House Business Books.

Reyes-Garcia, V., Babigumira, R., Pyhala, A., Wunder, S., Zorondo-Rodriguez, F., & Angelsen, A. (2016). Subjective wellbeing and income: Empirical patterns in the rural developing world. *Journal of Happiness Studies, 17*(2), 773–791. https://doi.org/10.1007/s10902-014-9608-2

Richardson, K., Steffen, W., Lucht, W., Bendtsen, J., Cornell, S. E., Donges, J. F., et al. (2023). Earth beyond six of nine planetary boundaries. *Science. Advances, 9*(37). https://doi.org/10.1126/sciadv.adh2458

Robeyns, I. (2017). Having too much. In J. Knight & M. Schwartzberg (Eds.), *Wealth, nomos LVIII* (pp. 1–44). New York University Press.

Römer, F., Henninger, J., & Le, T. D. (2021). International organizations and global labour standards. In K. Martens, D. Niemann, & A. Kaasch (Eds.), *International organizations in global social governance* (pp. 57–81). Palgrave Macmillan.

Ruggie, J. G. (1982). International regimes, transactions, and change: Embedded liberalism in the postwar economic order. *International Organization, 36*(2), 379–415. https://doi.org/10.1017/s0020818300018993

Ryan, R. M., Huta, V., & Deci, E. L. (2008). Living well: A self-determination theory perspective on eudaimonia. *Journal of Happiness Studies, 9*(1), 139–170. https://doi.org/10.1007/s10902-006-9023-4

Sabatier, P. A., & Jenkins-Smith, H. C. (Eds.). (1993). *Policy change and learning: An advocacy coalition approach*. Westview Press.

Sabato, S., & Mandelli, M. (2024). Towards an EU framework for a just transition: Welfare policies and politics for the socio-ecological transition. *European Political Science, 23*(1), 14–26. https://doi.org/10.1057/s41304-023-00458-1

Sakamoto, T. (2020). Social investment policy, economic growth, and welfare states: Channels of pro-growth effects of policy. *Social Forces, 99*(2), 590–615. https://doi.org/10.1093/sf/soz178

Sanne, C. (2002). Willing consumers—Or locked-in? Policies for a sustainable consumption. *Ecological Economics, 42*(1–2), 273–287. https://doi.org/10.1016/S0921-8009(02)00086-1

Schmelzer, M. (2016). *The hegemony of growth: The OECD and the making of growth*. Cambridge University Press.

Schneider, F., Kallis, G., & Martinez-Alier, J. (2010). Crisis or opportunity? Economic degrowth for social equity and ecological sustainability. Introduction to this special issue. *Journal of Cleaner Production, 18*(6), 511–518. https://doi.org/10.1016/j.jclepro.2010.01.014

Schor, J. B. (2004). *Born to buy: The commercialized child and the new consumer culture*. Scribner.

Schor, J. B. (2015). Work sharing. In G. D'Alisa, F. Demaria, & G. Kallis (Eds.), *Degrowth. A vocabulary for a new era* (pp. 195–198). Routledge.

Schröder, E., & Storm, S. (2020). Economic growth and carbon emissions: The road to 'hothouse earth' is paved with good intentions. *International Journal of Political Economy, 49*(2), 153–173. https://doi.org/10.1080/08911916.2020.1778866

Schulze Waltrup, R. (2024). The OECD, global challenges and contestation of the economic growth paradigm. In A. Isakova, M. Neuwinger, R. Schulze Waltrup, & O. Uraiqat (Eds.), *Constructing global challenges in world politics* (pp. 214–227). Routledge.

Schulze Waltrup, R. (2025a). An eco-social policy typology: From system reproduction to transformation. *Global Social Policy, 25*(1), 17–35. https://doi.org/10.1177/14680181231205777

Schulze Waltrup, R. (2025b). Questioning global social policy leitmotifs: Europe as a post-growth reference point amid planetary boundaries transgression. *Global Social Policy, 25*(1), 176–179. https://doi.org/10.1177/14680181241311994

Schulze Waltrup, R., Moore, M., & Paulsen, T. (2023). Eco-social policy in the global political economy: Analysing shifting discourses on agricultural subsidies. *European Journal of Social Security, 25*(4), 506–522. https://doi.org/10.1177/13882627231206002

Schulze Waltrup, R., Gengnagel, V., Zimmermann, K., & Kaasch, A. (2025). Setting the scene in Europe: Emerging national and transnational social policy responses to climate change. *Global Social Policy, 25*(1), 3–16. https://doi.org/10.1177/14680181241312557

Schumpeter, J. A. (1961 [1943]). *Capitalism, socialism and democracy*. George Allen and Unwin.

Sen, A. (1999). *Development as freedom*. Oxford University Press.

Seyfang, G., & Paavola, J. (2008). Inequality and sustainable consumption: Bridging the gaps. *Local Environment, 13*(8), 669–684. https://doi.org/10.1080/13549830802475559

Shandra, J. M., Shircliff, E., & London, B. (2011). World Bank lending and deforestation: A cross-national analysis. *International Sociology, 26*(3), 292–314. https://doi.org/10.1177/0268580910392260

Shove, E. (2014). Putting practice into policy: Reconfiguring questions of consumption and climate change. *Contemporary Social Science, 9*(4), 415–429. https://doi.org/10.1080/21582041.2012.692484

Skogstad, G., & Schmidt, V. A. (2017). Introduction: Policy paradigms, transnationalism, and domestic politics. In G. Skogstad & V. A. Schmidt (Eds.), *Policy paradigms, transnationalism, and domestic politics* (pp. 3–35). University of Toronto Press.

Smyth, P., & Deeming, C. (2018). Conclusion: Towards a new global social policy framework? In C. Deeming & P. Smyth (Eds.), *Reframing global social policy. Social investment for sustainable and inclusive growth* (pp. 323–342). Policy Press.

Sommer, J. M., Shandra, J. M., & Restivo, M. (2017). The World Bank, contradictory lending, and forests: A cross-national analysis of organized hypocrisy. *International Sociology, 32*(6), 707–730. https://doi.org/10.1177/0268580917722893

Soper, K. (2008). Alternative hedonism, cultural theory and the role of aesthetic revisioning. *Cultural Studies, 22*(5), 567–587. https://doi.org/10.1080/09502380802245829

Spangenberg, J. H. (2010). The growth discourse, growth policy and sustainable development: Two thought experiments. *Journal of Cleaner Production, 18*(6), 561–566. https://doi.org/10.1016/j.jclepro.2009.07.007

St. Clair, A. L., & Lawson, V. (2014). Poverty and climate change: The three tasks of transformative global social policy. In A. Kaasch & P. Stubbs (Eds.), *Transformations in global and regional social policies* (pp. 134–152). Palgrave Macmillan.

Stamm, I., Matthies, A.-L., Hirvilammi, T., & Närhi, K. (2020). Combining labour market and unemployment policies with environmental sustainability? A cross-national study on ecosocial innovations. *Journal of International and Comparative Social Policy, 36*(1), 42–56. https://doi.org/10.1017/ics.2020.4

Stern, P. C., & Aronson, E. (1984). *Energy use: The human dimension*. Freeman.

Stiglitz, J., Sen, A., & Fitoussi, J.-P. (2010). *Mis-measuring our lives. Why GDP doesn't add up*. The New Press.

Stone, D. (2003). The "Knowledge Bank" and the Global Development Network. *Global Governance: A Review of Multilateralism and International Organizations, 9*(1), 43–61.

Streeck, W., & Thelen, K. (Eds.). (2005). *Beyond continuity: Institutional change in advanced political economies*. Oxford University Press.

Surel, Y. (2000). The role of cognitive and normative frames in policy-making. *Journal of European Public Policy, 7*(4), 495–512. https://doi.org/10.1080/13501760050165334

Swenson, P. (2002). *Capitalists against markets: The making of labor markets and welfare states in the United States and Sweden*. Oxford University Press.

Tag, M. (2013). The cultural construction of global social policy: Theorizing formations and transformations. *Global Social Policy: An Interdisciplinary Journal of Public Policy and Social Development, 13*(1), 24–44. https://doi.org/10.1177/1468018112469805

Thaler, R. H., & Sunstein, C. R. (2008). *Nudge: Improving decisions about health, wealth and happiness*. Yale University Press.

Thomas, A. (2020). Reassessing the regressivity of the VAT. *OECD Taxation Working Papers*, (49), 0_1–51.

Thomas, A., & Doerflinger, N. (2020). Trade union strategies on climate change mitigation: Between opposition, hedging and support. *European Journal of Industrial Relations, 26*(4), 383–399. https://doi.org/10.1177/0959680120951700

Tienhaara, K. (2018). *Green Keynesianism and the global financial crisis.* Routledge.

Tienhaara, K., & Robinson, J. (2022). Introduction. In K. Tienhaara & J. Robinson (Eds.), *Routledge handbook on green new deal.* Routledge.

Tukker, A., Diaz Lopez, F. J., van de Lindt, M., Mont, O., Lorek, S., Spaneberg, J., & Giljum, S. (2009). *Sustainable consumption policies effectiveness evaluation (Scope2).* Report to European Commission within FP6.

Turnhout, E., Neves, K., & de Lijster, E. (2014). "Measurementality" in biodiversity governance: Knowledge, transparency, and the intergovernmental science-policy platform on biodiversity and ecosystem services (Ipbes). *Environment and Planning A: Economy and Space, 46*(3), 581–597. https://doi.org/10.1068/a4629

Unti, B. J. (2015). Job guarantee. In G. D'Alisa, F. Demaria, & G. Kallis (Eds.), *Degrowth. A vocabulary for a new era* (pp. 172–174). Routledge.

Utting, P. (2013). Pathways to sustainability in a crisis-ridden world. In R. Genevey, R. K. Pachauri, & L. Tubiana (Eds.), *Reducing inequalities: A sustainable development challenge* (pp. 175–190). TERI Press.

Vadén, T., Lähde, V., Majava, A., Järvensivu, P., Toivanen, T., & Eronen, J. T. (2020a). Raising the bar: On the type, size and timeline of a 'successful' decoupling. *Environmental Politics, 30*(3), 462–476. https://doi.org/10.1080/09644016.2020.1783951

Vadén, T., Lahde, V., Majava, A., Jarvensivu, P., Toivanen, T., Hakala, E., & Eronen, J. T. (2020b). Decoupling for ecological sustainability: A categorisation and review of research literature. *Environmental Science & Policy, 112*, 236–244. https://doi.org/10.1016/j.envsci.2020.06.016

Vetterlein, A. (2007). Economic growth, poverty reduction, and the role of social policies: The evolution of the World Bank's social development approach. *Global Governance: A Review of Multilateralism and International Organizations, 13*(4), 513–533. https://doi.org/10.1163/19426720-01304006

Vetterlein, A., & Schmidtke, T. (2024). The World Bank: A changing organization in a changing world. In A. Vetterlein & T. Schmidtke (Eds.), *The Elgar companion to the World Bank* (pp. 2–20). Edward Elgar Publishing.

Videira, N., Schneider, F., Sekulova, F., & Kallis, G. (2014). Improving understanding on degrowth pathways: An exploratory study using collaborative causal models. *Futures, 55*, 58–77. https://doi.org/10.1016/j.futures.2013.11.001

Von Gliszczynski, M. (2015). *Cash transfers and basic social protection.* Palgrave Macmillan.

Wade, R. H. (1996). Japan, the World Bank, and the art of paradigm maintenance: The East Asian Miracle in political perspective. *New Left Review I*, 2017.

Wade, R. H. (2002). US hegemony and the World Bank: The fight over people and ideas. *Review of International Political Economy, 9*(2), 215–243. https://doi.org/10.1080/09692290110126092

Wahlen, S., Heiskanen, E., & Aalto, K. (2012). Endorsing sustainable food consumption: Prospects from public catering. *Journal of Consumer Policy, 35*(1), 7–21. https://doi.org/10.1007/s10603-011-9183-4

Ward, J. D., Sutton, P. C., Werner, A. D., Costanza, R., Mohr, S. H., & Simmons, C. T. (2016). Is decoupling GDP growth from environmental impact possible? *PLoS One, 11*(10), e0164733. https://doi.org/10.1371/journal.pone.0164733

Waring, M. (1988). *If women counted: A new feminist economics*. Harper & Row.

WCED. (1987). *Our common future. World commission on environment and development*. Oxford University Press.

Wier, M., Birr-Pedersen, K., Jacobsen, H. K., & Klok, J. (2005). Are CO2 taxes regressive? Evidence from the Danish experience. *Ecological Economics, 52*(2), 239–251. https://doi.org/10.1016/j.ecolecon.2004.08.005

Wilkinson, R., & Pickett, K. (2009). *The spirit level. Why more equal societies almost always do better*. Allen Lane.

Winston, N. (2022). Sustainable community development: Integrating social and environmental sustainability for sustainable housing and communities. *Sustainable Development, 30*(1), 191–202. https://doi.org/10.1002/sd.2238

Woodward, R. (2009). *The Organization for Economic Co-operation and Development (OECD)*. Routledge.

World Bank. (1991). *The forest sector: A World Bank policy paper*. The World Bank.

World Bank. (1994). *Averting the old age crisis. Policies to protect the old and promote growth*. Oxford University Press.

World Bank. (1995). *Investing in people: The World Bank in action*. The World Bank.

World Bank. (2001). *World development report 2000/01: Attacking poverty*. Oxford University Press.

World Bank. (2005). *World development report 2006: Equity and development*. The World Bank.

World Bank. (2006). *World development report 2007: Development and the next generation*. The World Bank.

World Bank. (2010). *World development report 2010. Development and climate change*. The World Bank.

World Bank. (2012a). *Inclusive green growth. The pathway to sustainable development*. The World Bank.

World Bank. (2012b). *Toward a green, clean and resilient world for all. A World Bank Group environment strategy 2012–2022*. The World Bank.

World Bank. (2017). *Environmental and social framework*. The World Bank.

World Bank. (2019). *The World Bank Group action plan on climate change adaptation and resilience*. The World Bank.

World Bank. (2020a). *Resilient industries. Competitiveness in the face of disaster*. The World Bank.

World Bank. (2020b). *Transformative climate finance: A new approach for climate finance to achieve low-carbon resilient development in developing countries*. The World Bank.

World Bank. (2020c). *Developing a national green taxonomy. A World Bank guide*. The World Bank.

World Bank. (2021a). *The economic case for nature. A global earth-economy model to assess development pathways*. The World Bank.

World Bank. (2021b). *Climate change action plan 2021–2025. Supporting green, resilient and inclusive development*. The World Bank.

World Bank. (2021c). *Circular economy in industrial parks: Technologies for competitiveness*. The World Bank.

World Bank. (2021d). *State and trends of carbon pricing 2021*. The World Bank.

World Bank. (2021e). *Toolkits for policymakers to green the financial system*. The World Bank.

World Bank. (2022). *The role of extended producer responsibility scheme for packaging towards circular economies in APEC*. The World Bank.

World Bank & PMR. (2021). *A guide to developing domestic carbon crediting mechanisms*. The World Bank.

World Bank & UNEP. (2017). *Roadmap for a sustainable financial system*. The World Bank.

World Bank, & World Resource Institute. (2019). *Integrating green and gray. Creating next generation infrastructure*. The World Bank.

World Bank, Global Partnership for Financial Inclusion, & German Federal Ministry for Economic Cooperation and Development. (2017). *Financing for SMEs in sustainable global value chains*. The World Bank.

Zenghelis, D. (2012). *A strategy for restoring confidence and economic growth through green investment and innovation*. Retrieved from Grantham Research Institute on Climate Change and the Environment Centre for Climate Change Economics and Policy. (https://www.greenpolicyplatform.org/sites/default/files/downloads/resource/A_Strategy_for_Restoring_Confidence_and_Economic_Growth_CCCEP.pdf)

Zimmermann, K., & Graziano, P. (2020). Mapping different worlds of eco-welfare states. *Sustainability, 12*(5). https://doi.org/10.3390/su12051819

Index[1]

[1] Note: Page numbers followed by 'n' refer to notes.

R. Schulze Waltrup, *Eco-Social Policy in the OECD and the World Bank*, https://doi.org/10.1007/978-3-031-95429-0

Kolibri 360 GmbH
Ferdinand-Jühlke-Straße 7
99095 Erfurt, Deutschland
produktsicherheit@kolibri360.de